The Chromosomes and Their Disorders
An Introduction for Clinicians

D1429102

To BERYL CORNER

an eminent British paediatrician, whose example stimulated my interest in the disorders of children 40 years ago. For that continuing interest over the years, and for a most rewarding career, I offer her my grateful thanks.

The Chromosomes and Their Disorders
An Introduction for Clinicians

G. H. Valentine *MB ChB(Bristol) FRCP(Lond) FRCP(C) FCCMG*

Emeritus Professor of Paediatrics
University of Western Ontario

Clinical Medical Geneticist
Regional Birth Defects and Genetics Service,
Children's Hospital of Western Ontario
London, Canada

Fourth Edition

William Heinemann Medical Books
London

First published by William Heinemann Medical Books,
23 Bedford Square, London WC1B 3HH

First published 1966
Reprinted 1968
Translated into German 1968
Second Edition 1969
Reprinted 1971
Translated into Spanish 1971
Third Edition (reset) 1975
Fourth Edition 1986

ISBN 0 433 33603 X

M 9219
8/5/87

Typeset by Eta Services (Typesetters) Ltd., Beccles, Suffolk
and printed in Great Britain by BAS Printers Limited, Over Wallop, Hampshire

Contents

Non-disjunction
Mosaicism
Chimaerism
Causes of Chromosome Abnormalities
Nomenclature

Part II: Clinical Considerations

Preface to the Fourth Edition

Is there any need for this book at all, for this fourth edition? It might be argued that knowledge of the chromosomes and their disorders has become common currency, that chromosomes have become commonplace. But two things have happened since the last edition ten years ago that seem to justify another writing. Molecular genetics bids fair to revolutionise the lives of us all and to change medical practice out of all recognition, and a new 'audience' for a book such as this has arisen.

Part II of the book has not been greatly changed though almost all the karyotype illustrations are by a 'banding' technique and there is a new chapter on chromosomes and cancer. Part I has been completely rewritten to include the basics of molecular genetics and their implications. This exposition of this (to me at any rate) most difficult subject is, I know, very simplistic, but simplifications and half-truths can perhaps be excused when an effort is made to make the complex comprehensible. I should state here and now that until quite recently I have been a general paediatrician dealing with diarrhoea as much as with Down's syndrome and with croup as much as with chromosomes. This is my excuse.

Since the last edition genetics clinics have become much more numerous and widespread. The collection of data concerning the families to be counselled is done by genetics associates and nurse clinic coordinators. These indispensable workers may find something of value here.

As before I have chosen to be somewhat conversational in style, and I hope the reader will not be too much irritated by repetitions. I find I have to read new knowledge several times before I understand it. I suspect that others may feel as I do.

I make no apology for the lack of a bibliography that cites an authority for every statement made. Often I have forgotten where I have heard this or read that. This book is more to be regarded as genetical journalism than as a scientific treatise. It is, I hope, more than science fiction! In omitting acknowledgements I must invoke the prerogative of the journalist and state only that my facts come from 'informed sources'.

While I can only give thanks in general terms to all who may recognise their own work or thoughts, I must give special thanks to Dr Fred Sergovich and to his great team of dedicated cytogenetic technicians at the Cytogenetics Laboratory of the Victoria Hospital. Almost all of the karyotypes and several of the clinical illustrations were contributed by them. Without their help this new edition could never have been written. I must also express my gratitude to the Visual Arts Department of the Victoria Hospital for their skill in converting some of my scribbled drawings into comprehensible diagrams. I must also thank Dr J. Boone, Chairman of the Department of Paediatrics and Dr J. Jung, Chairman of the Division of Medical Genetics for their encouragement and for underwriting the costs of all the many illustrations.

My secretary, Vivian Huard, deserves great credit for her tenacity and perceptiveness in deciphering my illegible pencil-written manuscript and of correcting it into the finished text.

Lastly I would like to acknowledge a deep debt of gratitude to Beryl Corner who, by her enthusiasm and example, inspired me to become interested in disorders of children and, as a natural consequence, in genetics. When I graduated in Medicine during the Second World War, Beryl bore a very heavy burden of teaching and clinical responsibility, not the least of which was the grim occasion when, during an air-raid the Royal Hospital for Sick Children in Bristol had to be evacuated in the middle of the night. Despite so many concerns she found time to be interested in each student as an individual and to stimulate each one of us. On my retirement from clinical paediatrics and genetics I would like to thank Beryl for starting me on a most happy and rewarding career.

G.H.V.
1986

Genetics and Birth-defects Service,
Children's Hospital of Western Ontario
        and
The University of Western Ontario,
London, Canada

PART I

The Grammar of Cytogenetics

'Every problem becomes very childish when once it is explained to you. Here is an unexplained one. See what you can make of that, friend Watson.' I looked with amazement at the absurd hieroglyphics upon the paper.

'Why, Holmes, it is a child's drawing!' I cried.

'Oh, that's your idea!'

'What else should it be?'

'I think I can help you pass an hour in an interesting and profitable manner,' said Holmes, drawing up his chair and spreading out in front of him the various papers upon which were recorded the antics of the dancing men.

The Dancing Men
Arthur Conan Doyle

(The hieroglyphics above spell out, in the code of the dancing men, 'The Chromosomes and their Disorders'.)

CHAPTER 1

The Cell and its Chromosomes

As all mankind is believed by some to be descended from Adam and Eve, so all the innumerable cells of the body are descended from the first cell, the zygote, formed at the moment of conception by the union of the two gametes, the sperm and ovum. By myriads of multiplications through many generations that first cell and its descendants divide and divide again. As generation succeeds generation, the cells take different developmental lines, acquiring diverse forms and functions as they group themselves into masses and then into organs, each with its special purpose for the body as a whole.

Finally the body becomes a federation of organs and systems each with its teeming population of cell citizens who contribute to the common good and receive according to their several necessities.

The health and wellbeing of this federation of the body depends on the correct formation of the organs as their constituent cells divide and multiply; it depends on the correct functioning of those organs as determined by the disciplined and instructed activities of the cells of which they are composed. The demands of one organ must be met by the productive efforts of another. There should be no deficiencies nor surpluses. Each cell, each citizen, each province, each organ must work for the common good of the body as a whole. Here is the perfect economic community!

Not only must the confederation of the body be constructed and set in motion, but its form and its function must be maintained even though 50 million cell citizens die each second and 50 million are born anew to take their place. The torch must be passed from generation to generation as renewal repairs dilapidation.

Here in this book we will explore a group of clinical disorders in which quite major and visible disorders of structure of remote ancestral cells give rise to errors of form and function of the organs as generation of dividing cells succeeds generation in the same way that the present ills of mankind are, by some, attributed to the sin of Adam and Eve. First we must consider certain aspects of structure and function of the citizens of

3

the body, first given the name 'cells' by Robert Hooke three centuries ago.

The Cell

Whether it be from a man, fowl, fish or frog, whether it be from brain, liver, skin or muscle, there is a basic common structure to a cell. Upon this common anatomy are superimposed differences that suit a cell to its special function. Let us consider those essentials (Fig. 1).

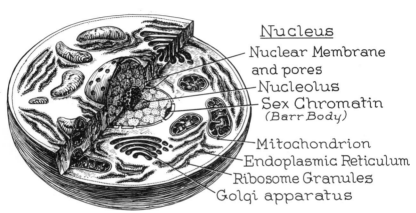

Nucleus

Nuclear Membrane and pores

Nucleolus

Sex Chromatin (Barr Body)

Mitochondrion

Endoplasmic Reticulum

Ribosome Granules

Golgi apparatus

Fig. 1. Schematic drawing of a typical cell showing its main component parts (adapted from The Cell, *Life Science Library, Time Inc., New York).*

There is the cell envelope which encloses its substance, more folded, invaginated and complex than in our illustration. (A motor nerve cell may have, as part of the cell envelope and its contents, an active filament, the neurone, several centimetres long.) Through the cell membrane can pass inward the requirements of the cell, but outward, also, its products.

The body of the cell, the cytoplasm, is traversed by a fine network of minute channels, the endoplasmic reticulum. Along these channels are clustered innumerable tiny granules, the ribosomes. Presumably this endoplasmic reticulum allows free passage within the cell of materials that, as we will see, are used by the ribosomes to forge the products of cell activity.

The mitochondria, roughly ovoid bodies divided internally by irregular membranes, are the site of oxidative metabolism; here is made, used and renewed adenosine triphosphate (ATP), the fuel of cell activity. The Golgi bodies, with their well-ordered arrangement of vacuoles serve as the warehouses and packing stations of cell products.

Somewhere within the cytoplasm lies the nucleus, contained within its perforated membrane, controlling all, but itself regulated by events both within the cell cytoplasm and in the cell's environment.

For a long time it has been known that every active living cell contains this nucleus, and that within this nucleus can be seen material that stains darkly with many stains. Except in the dividing cell this chromatin material appears as irregular shapeless masses with no pattern of arrangement. About a century ago it was observed that when a cell divides some order becomes apparent within the nucleus. The amorphous chromatin becomes condensed into a number of finite bodies, dimly recognisable in shape and in a number constant for the animal species. These are the chromosomes. These are the instructors of the cell, its board of governors, invisible when engaged in regulating the business of an active and functioning cell; visible when in recess, as the cell prepares and proceeds to divide.

The Chromosomes and DNA

Each chromosome is a long strand of deoxyribonucleic acid (DNA). In the larger chromosomes it is very, very long; in the smaller chromosomes it is still very long indeed. The DNA in all the chromosomes of each nucleus would, if fully extended, be several metres long—all that in the minute nucleus of a microscopic cell. Much condensation, coiling and looping is necessary to accommodate these long strands of DNA.

That the molecule of DNA was in the shape of a corkscrew or helix was first recognised by Rosalind Franklin in 1951. In 1953 Watson, Crick and Wilkins elucidated its fine structure and, in passing, remarked: 'It has not escaped our notice that the specific pairing we have postulated immediately suggests a possible copying mechanism for the genetic material.' Most prescient!

The DNA molecule is in the form of a double helix, a spiral ladder (Fig. 2) with the side pieces composed of a sugar, deoxyribose, and phosphate. The rungs of the ladder are made of four nucleotide bases: cytosine (C), guanine (G), adenine (A) and thymine (T). Two bases form each rung. Cytosine is always joined to guanine (CG or GC) and adenine likewise is always united to thymine (AT or TA) (Fig. 3). These four variants can succeed one another in any sequence. One could have a segment of ladder GC, CG, AT, TA or GC, CG, AT, AT or GC, TA, AT, GC, AT,

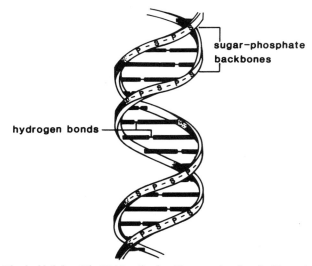

Fig. 2. The double helix of the DNA molecule with sugar-phosphate backbones joined by base-pairs.

CG. Innumerable formations and combinations are possible with a ladder of just a few rungs; 15 rungs can be arranged in more than a billion different ways. It has been estimated that the total human DNA ladder has, maybe, three billion rungs but much of it has no genetic function that we know. Much seems to be mere 'padding' between important sequences. One suspects that further research will show that nature is not so uselessly profligate and that these silent sequences have a real purpose. At least we can say that the number of possibilities of arrangement and rearrangement in the rungs of the ladder of the human chromosome complement are beyond comprehension. Perhaps 1 and 10 000 zeros to follow would scarcely suffice to count their number.

The length of the DNA ladder is measured, not in inches or centimetres, but in kilobases (kb): a thousand base-pairs or rungs is the unit of measurement. The unit of measurement of the distance between individual units of hereditary instruction, the genes, is given as centimorgans (cM).

How does one pack so much DNA into the chromosomes? How are they condensed at cell division? How are they extended in the active, the interphase cell?

First there is the helical structure of DNA itself; then the helix is wound around histone protein 'beads' to make a nucleosome (Fig. 4a).

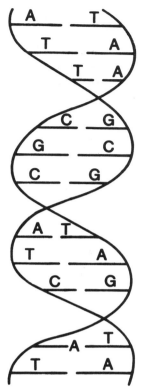

Fig. 3. *The base-pairs guanine–cytosine, cytosine–guanine, thymine–adenine and adenine–thymine joining the helical backbones.*

The nucleosomes themselves are joined as a strand which itself is coiled (Fig. 4b); that coil is then looped and re-looped into a mass that forms the body of the chromosome (Fig. 4c and d).

It is now well established that the sequences of rung arrangement give orders to the body cells as to the way they will be marshalled as the body develops, as to their function in the organs which they compose, as to the continuing activities of their descendants and their descendant's descendants, until death abruptly ends the activities of this amazing mechanism, uniquely created and set in motion at the moment of conception. How does it work?

Genes and Chromosomes

If we accept the statement of Watson and Crick that the specific pairing

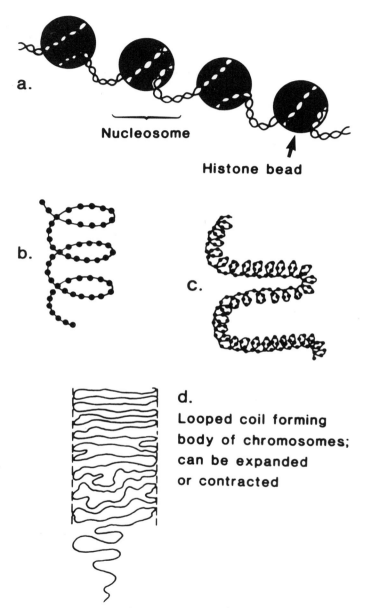

a.

Nucleosome

Histone bead

b.

c.

d.
**Looped coil forming
body of chromosomes;
can be expanded
or contracted**

Fig. 4. The microstructure of a chromosome.

of the nucleotide bases in the DNA molecule suggests a possible mechanism for the transmission of genetic information, we can imagine that segments of the DNA ladder (and the sequence of the rungs in that segment) might be concerned with individual cell activities; that a subdivision of the DNA strand might be concerned with the production of one particular product. And so it is.

There are specific segments or sequences concerned with specific activities and with the instructions to the cell as to how it will perform those activities. The unit of instruction is the gene. It is believed that each one of us, in every cell, has coded in the order of the four alternative rungs, in perhaps 100 000 meaningful sequences, precise instructions to the original first cell, the zygote (and to all its descendants through many generations) as to how it will divide, multiply and form organs with their many and diverse functions. Although each one of us is unique in the exact quality of all the instructions and cell activities, we have, of course, much in common with one another. It is in matters of detail that we differ within the limits of what we consider 'normal'. But a serious error of instruction in a matter of even a single detail can lead to a 'single gene disease', cystic fibrosis, for example. Groups of misinstructions or missed instructions, because many gene sequences together are over-represented or absent, are the subject of this book.

It is now well known that there are in each cell in humans 46 chromosomes, a rather large advisory council. They are in 23 pairs; both members of the pair and genes that they carry are concerned with the same matters of cell business, but not always seeing eye-to-eye as to just how that business is best conducted. Each member of the pair of similar, but not necessarily identical, instructions is one of a pair of homologous genes. They have the same place, locus, on the pair of homologous chromosomes. Alternative forms of the detail of a genetic instruction are called alleles. A pair of homologous genes, let us say on chromosomes number 9, may be concerned with the ABO blood group. One may instruct for blood group A, the other for group B. A and B are alleles at the ABO locus. In this instance the result is a person of blood group AB.

Of these 23 pairs of chromosomes, 22 are identical, concerned with the same matters of cell business in both males and females. These are the autosomal chromosomes or 'autosomes', 44 in all. The remaining pair, the sex chromosomes, are different in males and females. The female has a matching pair of sex chromosomes, XX. In the male, the pair do not match up. One of the male pair is congruent with an X chromosome but the other, the Y, is quite different in size, shape and in its genetic aptitudes. The female, then, is XX; the male XY.

The Y chromosome is very limited in its genetic capabilities. It has, so far as is known at present, one genetic preoccupation only—sex. It determines, as we will see, that its possessor develops as a male; without a Y in the chromosome complement the individual will develop as a female.

The X chromosome is concerned with many matters of business apart from sex differentiation. It is true that two X chromosomes are necessary for normal development of the ovaries, and that it has some genetic activities in assisting the Y chromosome to do its work in male sex determination, but the X chromosome can, in many respects, be regarded as an autosome. Over 100 gene loci have been assigned to the X chromosome.

Apart from its concerns with ovarian development and some part in male development, the X chromosome shows a unique peculiarity. Whereas both homologous genes on homologous autosomes exert their genetic effects, it is not so with the two X chromosomes and their paired genes.

In the very early fetus, perhaps up to 2 weeks from conception, both X chromosomes are active, but at that time, randomly, in each and every cell, one of the pair of X chromosomes becomes partly, but not entirely, inactive in exerting further genetic influence. In some cells one X is active; in others, the other. Thus, so far as the genes located on the X chromosome are concerned, some cells may get (if the genes are in disagreement) one instruction; others another. A female, then, is a genetic mixture, a 'mosaic', so far as genetic activity of the X chromosomes are concerned. This random inactivation of one X chromosome is known as lyonisation, after Mary Lyon who first hypothesised its happening.

The nuclear chromosomes are not the only genetic instructors of the cell. In the mitochondria of the cell there is also DNA of some 16 000 base-pairs concerned with 37 separate instructions. Little seems to be known about any clinical effects of this mitochondrial DNA, but it may have some relevance to those genetic disorders, neurofibromatosis and myotonic dystrophy for example, in which the genetic disease is more severe when inherited from the mother than from the father. It may be that mitochondrial genes, contained in the cytoplasm of the ovum, but not in the sperm, influence the effect of the nuclear chromosomal genes.

Alleles

While the gene is the unit of genetic instruction, it may have alternatives in matters of detail; one gene is brought to the zygote in the sperm, the other in the ovum. As to the instructions that the cell receives, the alleles may entirely agree; the individual is homozygous for that genetic affair. Or they may disagree; the individual will be heterozygous for that genetic instruction. Amicable agreement to disagree will allow the effects of both disparate genes to show themselves, to be expressed. A father of blood group A and mother group B can have a child who equally expresses the group of each: AB. This is co-dominance. But with heterozygosity for an instruction, there can be unequal disagreement; one can be stronger and dominant to the weaker recessive allele, the alternative instruction. The allele for neurofibromatosis is stronger than the allele for normality. It is dominant. The possessor of one allele for normality and one for neurofibromatosis will show the disease. On the other hand the allele that misinstructs for the production of the enzyme phenylalanine hydroxylase is recessive to the normal allele. The heterozygote carrier of the abnormal allele will be normal phenotypically: normal, that is, for everyday practical purposes. The normal allele will take precedence. Only if, by unlucky chance or consanguinity, two heterozygotes of a recessive allele mate will there be abnormal children. If there should be such a mating, any child would run a 25 per cent risk of receiving a double dose of the abnormal allele; that homozygosity for the abnormal recessive alleles would cause phenylketonuria (PKU).

If an abnormal and recessive allele is represented on only one of the two X chromosomes of an XX female, no phenotypic effect is shown. The normal outweighs the abnormal. There is rarely any expression of an X-linked recessive allele in females. In an XY male it is otherwise. His Y chromosome has no normal allele to overcome the abnormal one. He will show a phenotypic effect, be it haemophilia, Duchenne muscular dystrophy or one of many other X-linked recessive genes. He is hemizygous.

Suppose that, by chance, in the carrier of an abnormal X-linked recessive allele there is an unequal inactivation of the X chromosomes, unequal lyonisation, so that, instead of roughly 50:50, there were to be a 40:60 ratio; there could be a predominance of cells with the X chromosome bearing the abnormal allele active in the body's population. Sometimes in this way heterozygous females can show more or less features of an X-linked recessive disease.

Or suppose that an X-chromosome were to be missing; suppose an individual had 45 chromosomes only, rather than 46. That person could show the ill-effects of a single abnormal allele. A 45, single-X, female could have haemophilia in the same way as could a hemizygous male.

Genetic instructions have their times and places; not every cell does everything all the time. At points along the way in fetal development genetic instructions dictate cell differentiation, cell proliferation and organ formation. The liver forms, the skin develops. As time goes on, the liver cells are instructed as to what liver cells should do—similarly with skin. Although every cell in every organ contains the whole genome, the whole genetic encyclopaedia, not all genes are active in all tissues. The gene instructing for the enzyme phenylalanine hydroxylase is active in the liver, but not in the skin. A study of fetal skin cells, obtained in pregnancy by amniocentesis, cannot tell you, by estimating the enzyme activity of the fetal cells, if the fetus has the disease PKU; normally there will be no enzyme activity, no evidence of that genetic instruction. In skin cells, the gene for that enzyme is normally 'switched off'. On the other hand, the gene for hexoseaminidase is normally switched on in skin cells. Fetal skin cells can be used for the prenatal diagnosis of Tay Sachs disease.

There is also regulation in time. A main plasma protein of the early fetus is alphafetoprotein (AFP). It is manufactured under genetic instruction. As the fetus matures, the gene for AFP is switched off and the genes for albumen and globulin are switched on. In the fetus, the haemoglobin is haemoglobin, F, HbF. In early infancy, HbF is phased out; adult haemoglobin HbA is switched on in its stead. The single dominant allele that determines Huntington's chorea, for instance, only becomes manifest, although present all along, in middle life. Genes have temporal regulation.

Genetic activity can be induced by necessity. The enzymes catabolising bilirubin, or alcohol, are made in greater quantity if a greater quantity of those enzymes is required, up to a certain limit. Phenobarbitone induces increased production of several enzymes. Many, perhaps most drugs, act by regulation of gene transcription and translation. Genes can be regulated by environmental events.

The Genetic Code, Transcription

Suppose we have a segment on one of the chromosomes with a sequence of rungs thus AT, TA, TA, CG, GC, CG, AT, TA, CG, AT, TA, and so on, we can see that this could be a code. It is.

In the cell, busy about its everyday metabolic affairs, the ladders of sequences in the cell nucleus are enormously extended and all the chromosomes, except for the one partly inactivated X chromosome in females, are in a long filamentous form. But each of these long filaments of DNA is in a constant state of cleavage and reformation, the breakages being where the two halves of each rung are joined by a hydrogen bond. One half of the ladder would run A, T, T, C, G, C, A, T, C, A, T, the other T, A, A, G, C, G, T, A, G, T, A.

Alongside the broken ladder, split open by an enzyme, RNA polymerase, a sugar-phosphate stem, ribonucleic acid, moves in. Onto this stem new rungs are built to correspond with the rungs of the ladder that has broken away. There is complementary transcription (Fig. 5) as mRNA, messenger RNA, is built up.

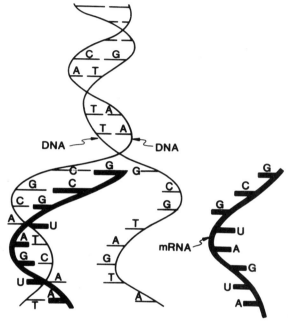

Fig. 5. *Transcription of the genetic code from DNA to mRNA.*

The mRNA is not identically complementary. The reciprocal of A is not thymine, T, but another base uracil, U. The sequence T, C, G, C, A, T, C, A, T induces on the new stem the sequence A, G, C, G, U, A, G, U, A. We now have, with the exception of the substitution of U for T, a reproduction of half of the old ladder. New ladders, built up on the template of the sequences of the base-pairs in the DNA molecule are the copies of the genetic material, as suggested by Watson and Crick: 'It has not escaped our notice that the specific pairing we have postulated immediately suggests a possible copying mechanism for the genetic material.'

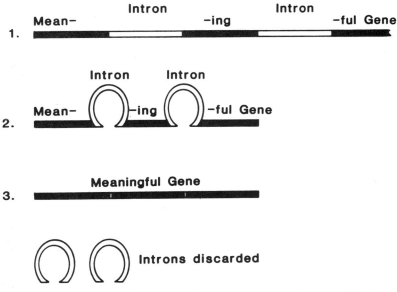

Fig. 6. *Removal of the redundant introns before the mRNA can be translated by the ribosomes.*

Translation

When a protein food is digested it is split down, first to peptones, then to polypeptides and then to the basic building blocks of proteins, the amino acids. There are 20 different amino acids which, in varying combinations and quantities make up the very many proteins and enzymes of the body. Amino acids absorbed into the body bathe all the cells and enter their cytoplasm. The amino acids from, let us say, a beef steak, enter the cells of man.

Before the messenger RNA, mRNA, leaves the nucleus on its way to the ribosomes, much DNA 'junk' must be removed. For reasons that are not understood the DNA strands contain much useless material between meaningful gene sequences and even within the length of a gene sequence. Coding sequences, exons, are mixed up with, and separated by, apparently useless introns (Fig. 6). Perhaps these introns served our remote ancestors in some way; perhaps they are merely irrelevant remnants of the DNA of viruses that invaded the nuclear DNA of our ancestors in remote times past; or their usefulness may yet be discovered.

The mRNA, stripped of its useless introns, leaves the cell nucleus and travels to groups of ribosomes, granular masses composed of protein and another kind of RNA (which need not bother us). There, among the ribosomes, which act as anvils upon which the code of the base-sequences is translated, the amino acids are welded together in the correct order that will build up a specific polypeptide that will be further built into proteins and enzymes.

The long strand of mRNA, perhaps 1000 bases long for a single gene, is divided into small units of activity: three bases form a unit, a 'codon'. GAC might be a codon; so might GGU, AGA, CAU, AUC, or UUU. The four bases cytosine, guanine, adenine and uracil, in groups of three, can be arranged in 64 different ways (Fig. 7). Since there are only 20 amino acids relevant to human use, some of the codons of this 'triplet code' are redundant. A single amino acid may have more than one codon. For example, the amino acid, valine, may be coded by the codons GUU, GUC, GUA, or GUC, whereas tyrosine is coded for by uracil–adenine–uracil or by uracil–adenine–cytosine: UAU and UAC. The sequence given above as an example of transcription, A, G, C, G, U, A, G, U, A, codes for serine, valine and valine. What do we mean by 'coded by' and 'put in correct order'?

In the cell cytoplasm, and available for synthesis into proteins and enzymes, are the amino acids. Each has an affinity for, and is attracted by, its own specific codon—but not directly; these are intermediaries: transfer RNA (tRNA). These tRNA intermediaries have a cloverleaf configuration and each has three unpaired base molecules projecting from it at one end and a site for attachment of an amino acid at the other (Fig. 8). The three bases of each tRNA are an 'anticodon', complementary to the codon of mRNA for each amino acid. If the mRNA codon for lysine is (see Fig. 7) either AAA or AAG, the tRNA to which lysine would be attached would be either UUU or UUC. A sequence of mRNA triplet codes would attract, in order, the amino acids anchored to the complementary tRNA anticodons (Fig. 9).

The mRNA, shorn of its introns, comes to the ribosomes in the cytoplasm. Here it passes between two parts of a ribosome—somewhat in the same way as clothes go through a wringer—and, one by one, as each codon of mRNA passes through, the complementary tRNA and its attached amino acid is summoned from the amino acid pool and the amino acids, in sequence dictated by mRNA, are welded into a specific product. Punctuation marks signalling the start of instructions for one product and the beginning of the next are given by the codons AUG and UAG. The tRNA, liberated from its amino acid, is available to be used again.

Abbreviation	Amino Acid
ala	alanine
arg	arginine
asn	asparagine
asp	aspartic acid
cys	cysteine
gln	glutamine
glu	glutamic acid
gly	glycine
his	histidine
ile	isoleucine
leu	leucine
lys	lysine
met	methionine
phe	phenylalanine
pro	proline
ser	serine
thr	threonine
trp	tryptophan
tyr	tyrosine
val	valine

U			
UUU } PHE UUC UUA } LEU UUG	UCU } SER UCC UCA UCG	UAU } TYR UAC UAA } STOP UAG	UGU } CYS UGC UGA STOP UGG TRP
C			
CUU } LEU CUC CUA CUG	CCU } PRO CCC CCA CCG	CAU } HIS CAC CAA } GLP CAG	CGU } ARG CGC CGA CGG
A			
AUU } ILE AUC AUA AUG MET	ACU } THR ACC ACA ACG	AAU } ASN AAC AAA } LYS AAG	AGU } SER AGC AGA } ARG AGG
G			
GUU } VAL GUC GUA GUG	GCU } ALA GCC GCA GCG	GAU } ASP GAC GAA } GLU GAG	GGU } GLY GGC GGA GGG

Fig. 7. The genetic code, each amino acid is coded for by one or more triplet codons.

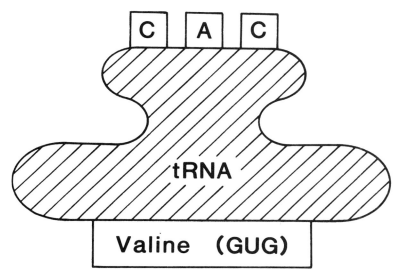

Fig. 8. Transfer RNA, tRNA, with its specific complementary codon selecting its specific amino acid to add to the polypeptide chain.

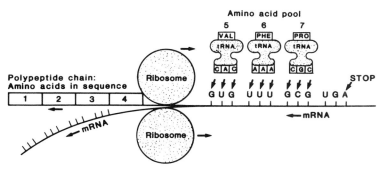

Fig. 9. Translation of the genetic code and the sequential assembly of the amino acids to form the polypeptide chain.

Gene Regulation

As we have said, although every cell of the body contains the whole enormous spectrum of genes, the genome, they are not all active in all cells at all times. At the very first, genes concerned with cell division, multiplication and differentiation are the most active. As organs take shape and assume their specific functions, genes coding for enzymes and proteins become 'switched on', while the genes relating to cell division and proliferation become switched off and remain quiescent unless injury and repair require their reactivation. Sometimes unrestrained and unrequited reactivation gives rise to useless cell proliferation—cancers.

Some genes instructing for enzymes and proteins, structural genes, have but a brief span of activity. The genes coding for alphafetoprotein and haemoglobin F have, normally, almost ceased their activities by the time of birth. Others, of course, continue their activities, with varying degrees of enthusiasm, throughout life, directing the innumerable activities of the body as it adapts to environmental circumstances.

While genes instruct, via the DNA molecule, mRNA and its codons, tRNA and its anticodons, as to what products will be forged by the ribosomes, the genes themselves are not autonomous; they are regulated by environmental necessities.

For example, in intrauterine life bilirubin is being produced in the fetus. Before birth it is eliminated through the placenta into the mother's circulation. The fetus has no need to deal with bilirubin for itself. At birth, it is on its own. In this new environmental situation, bilirubin starts to accumulate; almost all newborn babies have a rising serum bilirubin level and the liver cells become exposed to this new situation. Signals are sent to the nuclear DNA requesting, if you like, the transcription of the gene coding for the enzyme glucuronyl transferase. The gene is transcribed and in due course translated into the assembly of the polypeptides which go to make up the enzyme which joins glucuronic acid to bilirubin and, in so doing, makes it water-soluble, and excretable in the bile. Genes can be turned on by a necessity for their product; or they can be turned on by other, perhaps physiologically quite irrelevant, means. Phenobarbitone can help to switch on the gene coding for the glucuronyl transferase enzyme. Many drugs, many hormones, indeed perhaps all environmental factors that result in altered adaptive behaviour do so through influencing gene transcription or translation. There is two-way communication between the nucleus and the cell cytoplasm.

The precise mechanics of gene regulation are poorly understood but surely transcription of DNA onto mRNA must be regulated by the opening up, or otherwise, of the double helix at the base-pair bonds to greater or less degree and for varying durations. Presumably there is some regulation of the degree to which the enzyme RNA–polymerase can move in and build up mRNA. One suggested mechanism can be simplified as follows.

Genes seem to work in clusters concerned with much the same activi-

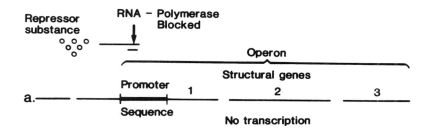

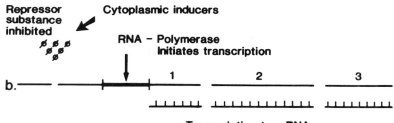

Fig. 10. Gene regulation, a simple mechanism by which gene activity might be regulated by environmental circumstances.

ties—interspersed, no doubt, with introns. Their concerted activities are controlled by an operator or promoter sequence. The unit, promoter and structural (coding for a product) genes is an operon. Unless required by circumstances, that operon is non-functional; the relevant part of the DNA molecule is not opened up. The operon is repressed by a repressor substance that keeps the molecule closed and inhibits RNA–polymerase. When cytoplasmic influences, whatever they may be, require the operon to be switched on, the repressor substance is inactivated, RNA–polymerase moves in, the DNA splits open and the code is transcribed to mRNA (Fig. 10).

Apart from cytoplasmic influences, genes act upon one another, assisting or inhibiting. Base-pair sequences can jump from one place to another along the DNA strand and such a 'jumping gene' can become interposed between a promoter sequence and its structural genes; or a promoter sequence can become detached from the structural genes and itself may jump elsewhere. It seems that DNA is less rigid than we have thought. It may well be that 'jumping genes' and inhibiting transposed sequences are responsible for 'non-penetrance' of a structural gene: no functional result, no product made, no phenotypic expression, even when one knows from pedigree analysis that the gene must be present in the genome.

Gene Identification: the 'New Genetics'

Until about 5 years ago the presence of a gene in the genome of an individual was inferred by recognition of the activities of that gene—the phenotypic result of the activities of the gene. To be sure, one sometimes was able, by study of the family pedigree, to deduce that a gene must be present even though there was no effect in a particular individual, even though the gene was 'non-penetrant'. But one was dependent on observing the phenotypic effects in other family members. It might be said that 'the old genetics' was 'phenotypic genetics'. All that has changed and continues to change explosively. The 'new genetics' can identify the presence of the gene itself, before it may ever show itself in the phenotype, and even in cells of tissues that are never affected by the presence of the gene. Perhaps the 'new genetics' can be called 'genotypic genetics'.

As we have seen, a gene is a segment of DNA with maybe 1000 complementary and reciprocal base-pairs making up a specific gene sequence. We have also seen that the DNA double helix can be split apart into single strands and that, as mRNA is built up, each base attracts to the RNA strand its complementary base. A sequence of bases on single-stranded DNA has an affinity with a complementary strand of bases. If one were able to obtain, or make, a single strand complementary to the single strand of a gene sequence, it is not difficult to imagine that, introduced to that gene sequence, it might 'lock' onto it, even as in the natural state two strands and their reciprocal bases are locked together.

Now suppose it were possible to label the strand that you had obtained or made with radioactivity or some other means for its detection, you could be fairly said to have a 'probe' to detect a gene sequence whose presence or absence is in question. If you were able to show that your probe had locked onto a segment of DNA, you could fairly say that your probe had detected its complementary gene sequence.

It is possible to cut up the enormously long strands of DNA that comprise the chromosomes; they can be chopped into fragments by enzymes called 'restriction endonucleases'. (They are so called because they can chop up and destroy the DNA of a virus that can invade and infect the bacteria from which they are obtained. They 'restrict' the effects of virus infections upon bacteria.) These restriction endonucleases, of which there are dozens, have strange names: BamHIII, EcoRI, HindIII, MstII, XbaI, to name but a few. They all have a common property: to 'recognise' a specific sequence in a DNA strand and to cut the strand at that point of recognition. Figure 11 shows how HaeIII cuts directly across the sequence GGCC and its reciprocal CCGG and how BamHI cuts at two non-opposed sequences leaving 'tailed' strands.

Hae**III**   BamHI

GG|CC   **G|GATC C**
CC|GG   **C CTAG|G**

Fig. 11. Cutting the DNA molecule by restriction endonuclease enzymes.

The total DNA, the genomic DNA, can be extracted from cells mechanically and chemically. Any cells can be used; they all contain the same DNA. A minute quantity, just a few micrograms (μg) of DNA is all that is needed for a test.

This DNA can now be treated with one or more restriction endonucleases, chopping it into fragments of different lengths wherever the sequences cleaved by the enzyme are recognised. With one enzyme one will get one set of fragments, with another one will get different fragments. Using several enzymes, the DNA can be cut up into a collection of very many bits and pieces: a so-called 'DNA library', a veritable treasury of genetic information. Because we are all genetically unique, the library of each one of us is unique—even though we do have many genes in common.

The mixture of DNA fragments are run electrophoretically on an agarose gel plate with the digests of the different endonucleases being run side by side. The smaller fragments will move furthest from the − to the + pole in the gel; the fragments will be stratified and separated by size and by molecular weight.

For convenience of handling and permanence, the DNA fragments in the agarose gel are 'Southern (the inventor of the process) blotted' onto a cellulose nitrate sheet by simple mechanical squashing together of the gel and sheet. The DNA, stratified as before, is transferred to the sheet. Next, the DNA is split into single strands, halves of the double helix, with their sequences of bases ready to be tested with a probe specific for the gene that is sought. How might one make a probe?

If one knows the chemical composition and the molecular structure of a gene product, one may know the sequence of the amino acids of which it is made. Knowing the codons for each amino acid one can deduce the sequence of bases in the mRNA. One can then assemble an mRNA strand, or at least enough of it to be specifically and uniquely identical with the mRNA that is transcribed by the gene under investigation. This synthetic mRNA is then treated with an enzyme, reverse transcriptase, that reproduces a segment of DNA, cDNA, complementary to the mRNA. By a process called 'nick-translation' (so called because an enzyme DNaseI makes nicks in the cDNA molecule), this cDNA is made radioactive by inserting P^{32} into the nicks (or it can be labelled in other ways) so that its presence can be detected.

The nitrocellulose sheet and its absorbed DNA fragments are incubated with the probe. If a DNA sequence complementary to the probe is present on the sheet, it will lock into, or 'hybridise' to, the probe. A stratum, a fragment, on the sheet will show detectable radioactivity. A specific gene sequence has been located in the genome—irrespective of its phenotypic activities, if any, in the cells from which the DNA was derived (Fig. 12).

Perhaps a more natural way to make a probe would be to extract from the relevant cells, reticulocytes in the case of haemoglobin synthesising genes, the relevant mRNA. This mRNA can be converted as with the synthetic mRNA, to cDNA using, again, reverse transcriptase. The yield of cDNA will be very small, but the amount can be increased by 'cloning', by 'recombinant DNA' technology. Figure 13 shows one way to do it.

Bacteria, *Escherichia coli*, have in their bodies strands of DNA in the form of rings. These are 'plasmids'. Using a restriction endonuclease, these plasmid rings can be cut and opened up. The cDNA to be multiplied by cloning can be cut with the same enzyme and, using another enzyme, 'ligase', it can be inserted and welded into the ring of plasmid DNA; a 'new' DNA has been created by combination with another DNA; hence the term 'recombinant DNA'. The bacteria can, of course, be enormously multiplied by culture as will be their contained plasmids and the inserted DNA. This much augmented DNA can be cut out from the multiplied bacterial plasmids, labelled by nick translation and used as a hybridising probe. There are other ways by which probes can be made and cloned, but the above will perhaps more than suffice. At any rate, specific sequences of DNA can be identified using DNA probes.

In the same way that normal genes can be located by DNA probes, abnormal genes with their abnormal sequences can be located among the DNA fragments in the genomic library. To date (the summer of 1985),

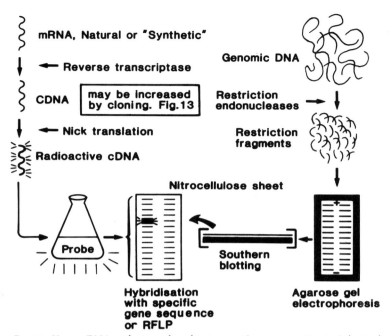

Fig. 12. How a DNA probe is used to detect a specific sequence cut out of the total genomic DNA by restriction endonucleases.

rather few of the very many abnormal genes can be specifically picked out. The gene for sickle-cell haemoglobin and sickle-cell disease can be directly identified and so can the genes for certain thalassaemias; so also can the genes coding for phenylalanine hydroxylase and phenylketo-nuria, hypoxanthine ribosyltransferase and Lesch–Nyhan syndrome, ornithine transcarbamylase and arginosuccinate synthetase which give urea cycle disorders with hyperammonaemia and alpha-1-antitrypsin and the disorder of its deficiency. Most excitingly, the genes for the co-agulation factors VIII and IX have been cloned and specific probes can be used to detect, even before birth, the presence of the genes for haemo-philia and Christmas disease. The list of specific probes is rapidly expanding but we sorely lack at this present time probes that will pick out the genes for cystic fibrosis and Duchenne muscular dystrophy. One expects to have these probes within a year or two.

Even if one lacks specific probes, one may learn by inference and deduction whether a gene is likely to be present or not. It works this way. Suppose there are two boys, Tom and Dick, who are inseparable

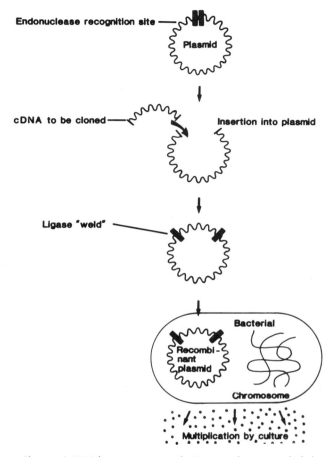

Fig. 13. *Cloning of cDNA by inserting it into the DNA of a bacterium which, by its own multiplication, will multiply the cDNA.*

friends, linked, one might say, by such a close bond of friendship that they almost invariably go around together; suppose that they go out into the bush, the forest, Tom wearing a bright red jacket, Dick a green camouflage coat; suppose that they fail to come home and search parties are sent out. It might be difficult indeed to find Dick in his green coat, but if you can find Tom by his bright red jacket, one could be reasonably certain that his great friend, Dick, is close at hand; provided, that is, that they have not become separated by some accident.

One may not have a specific probe for an important gene, but one might have one for a DNA sequence so closely linked that the detection of one infers the presence of the other but with an important qualification: that the friendship, the linkage, must be so close on the chromosome (so few centimorgans apart) that the testing sequence and the looked-for gene are most unlikely to become separated. The closer the linkage, the more reliable the test. The closer the bond, the greater the chance of finding the errant Dick. Of course, if Dick were to be 'flanked' on each side by two friends, Tom and Harry, one might be even more certain to find him. He is less likely to become separated from both of them.

As we have noted there is much redundant DNA between structural, product producing, gene sequences. There will be sequences, not necessarily in any way related to the function of the gene in which you are interested, that closely flank that gene. These flanking sequences are called 'restriction fragment length polymorphisms', RFLPs for short. If closely enough linked they will be inherited, without separation, along with the gene that they flank. They can be used as 'markers' for an abnormal gene. Here is how this works.

We do not all of us have exactly the same RFLP sequences flanking our structural genes. There can be detectable differences. Let us take a simple example. Suppose we have a man who has a hereditary disease that shows itself in the heterozygote (one 'good' gene, one 'bad') and that this disease, let us say it is Huntington's chorea, does not show itself until mid-adult life, we might, for one reason or another, wish to know whether his child (or even his unborn child) had received the 'bad' gene or not. He has a sister with the disease, but an unaffected brother. His father had the disease but both parents are dead. His wife is unaffected (Fig. 14).

Suppose there is a 'marker RFLP' very, very close to the locus of the Huntington's chorea gene (as indeed there is) with three distinguishable alternatives, A, B and C. We might determine that our patient has both A and B marker sequences. His affected sister has the same markers, A, B. In the normal brother, only A sequences can be detected; he is AA. The wife of our patient has no A nor B sequence, she is CC. What can we say of the child who is CB? He has a marker from one parent, one from the other; the C must have come from mother, B from father. The normal brother is AA, but both affected persons are AB. The B sequence must in this family (but not necessarily in all families with Huntington's chorea in their pedigrees) be linked to the Huntington's chorea gene. So one can say that, provided there has been no separation of the marker sequence from the Huntington's chorea gene in the meiotic cell division that has

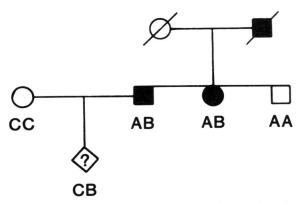

Fig. 14. How a marker restriction fragment length polymorphism can be used to infer the presence or absence of a relevant gene.

made the sperm that conceived the child (that is to say, provided the gene linkage was close enough to make that separation extremely unlikely), the CB child must have inherited the Huntington's chorea gene. If the father had been homozygous for the B marker, BB, the family would have been 'uninformative'; we could not know whether the B marker in his child is linked to the Huntington's chorea gene or to its normal counterpart.

Such linkage studies using DNA probes to identify marker RFLP sequences can be used to detect such otherwise undetectable genes as those for retinitis pigmentosa and Huntington's chorea. While there are detectable flanking RFLPs on either side of the Duchenne dystrophy gene, the linkage is not close enough that one can with certainty ascertain the presence or otherwise of the Duchenne dystrophy gene, though one may be able to deduce if a woman at risk is likely to be a carrier of the gene, or not.

By now, I am sure, the reader must wonder if this is a book about chromosomes or molecular genetics. It is a good question! But one cannot divorce the one from the other. DNA is the stuff of chromosomes; base-pairs and their sequences are the stuff of DNA. Moreover, nowadays, no student of genetics and chromosomes, even at a most elementary level can avoid the jargon of molecular genetics for we are entering a new era of 'predictive genetics' which will lead to a new era of 'predictive medicine', the recognition of those who are genetically at especial risk of certain sicknesses. This author, clinician though he is, has had to come to grips with at least this much of molecular genetics. In the section on prenatal diagnosis we will briefly revert to the subject of DNA sequence analysis.

And now to the visible structure of the human chromosomes and how they can be seen.

Recommended Further Reading

General Genetics

Conick, Larry and Wheelis, Mark (1983). *The Cartoon Guide to Genetics.* Barnes and Noble Books, New York, Cambridge etc.

Mange, Arthur and Mange, Elaine (1980). *Genetics: Human Aspects.* Saunders Co., Philadelphia.

Thompson, J. and Thompson, M. (1980). *Genetics in Medicine.* W. B. Saunders Co., Philadelphia.

Molecular Genetics

Emery, Alan E. H. (1980). *An Introduction to Recombinant DNA.* John Wiley, Chichester and New York.

DNA in medicine (1984). *Lancet* series of ten weekly articles, commencing 13 October 84 through to 22 December 1984; especially those of 13 October, 20 October, 27 October, 17 November, 1 December, 15 December, 22 December. *Lancet*, London.

CHAPTER 2

The Human Chromosome Complement

The Barr Body

Although it has few clinical applications today, the Barr body deserves recognition and description. Here in London, Ontario, 36 years ago, Murray Barr, a neuroanatomist and his graduate student Ewart Bartram recognised that among the jumble of chromatin material in the nucleus, a darkly staining oval body could often be seen alongside the nuclear membrane in the nerve cells of the cats which were the objects of their investigations into adaptation to high-altitude flight (Fig. 15). Often they

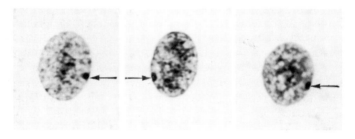

Fig. 15. The Barr body, partial inactivation of one X chromosome.

could be seen, but not always. They then realised that when this chromatin body could be seen, the cats were female. The body was not seen in tomcats. This certainly was a serendipitous observation! In their original publication they stated 'The sex of a somatic cell ... may be distinguished with no more elaborate equipment than a compound microscope'.

It was very soon realised that this Barr body represented one of the X chromosomes. As we have already noted, although the zygote and all its early descendants have both X chromosomes active about their genetic business, at about 2 weeks after conception, quite randomly among the cell population, in one cell one X chromosome will be partially inactivated; in another, the other. It is this process of X-inactivation, this lyonisation, that forms the Barr body. The inactive X becomes condensed and visible. Thereafter all descendants of each cell perpetuate the X-inactivation of their forebears so that, in females, about 50 per cent of cells have one X active and about 50 per cent the other. In XY males, of course, there is only one X and it must be active in all cells for the embryo to survive. It is clear that the inactivation of one X can only be partial; the inactive X cannot be entirely without genetic functions. There are females who have as a chromosome abnormality only one X. If one of two Xs were entirely superfluous they would be entirely normal. As we will see later in this book, they are not.

The effect of this X-inactivation and its resulting populations of disparate cells can sometimes be observed clinically. If a woman is heterozygous for the X-linked genetic disorder anhydrotic ectodermal dysplasia, that is to say one X chromosome carries the abnormal gene and the other the normal, the result of the discordant instructions to the cells can be seen. One patch of skin can sweat; another cannot. In women heterozygous for the X-linked gene for the red cell enzyme defect glucose-6-phosphate dehydrogenase (G6PD) deficiency two red cell populations can be recognised: those that are normal and those that are not.

The X inactivation is quantitative. One X, no inactivation, no Barr body. Two XXs, whatever else there may be, if any, in the way of Ys, one X is inactivated, and there is one Barr body; three Xs, two Barr bodies, four Xs, three (Fig. 16). There is, then, an n-1 rule. The number of X chromosomes is one more than the number of Barr bodies (Fig. 17).

This X-inactivation can have some diagnostic value. As we have noted earlier, two X chromosomes are required for normal ovarian development. Let us suppose that an adolescent girl has no breast develop-

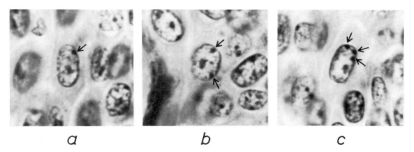

a   b   c

Fig. 16. *One Barr body, two X chromosomes, two Barr bodies three X chromosomes, three Barr bodies four X chromosomes.*

(a) *no* *sex chromatin*	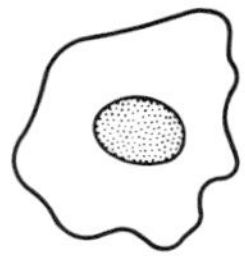	XO, XY, XYY.
(b) *single* *sex chromatin*	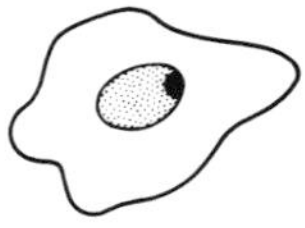	XX, XXY, XXYY.
(c) *two masses of* *sex chromatin*		XXX, XXXY, XXXYY.
(d) *three masses of* *sex chromatin*		XXXX, XXXXY.
(e) *four masses of* *sex chromatin*		XXXXX.

Fig. 17. The n-1 rule.

ment and has not started menstruation, it could be because she has little or no ovarian development. If she had no Barr bodies in her cells that would become almost diagnostic certainty. How could one look at this? Very simply.

Cells are scraped with a blunt spatula from the mucosal surface of the mouth, the buccal mucosa. The cells are smeared onto a microscope slide, fixed, stained and examined under the microscope for the presence or absence of Barr bodies.

Let us suppose that if a newborn baby has quite ambiguous genitalia one might learn within a very short time something about the 'genetic sex' of the baby: what it might have been destined to have become, boy or girl, if all had gone right in development; but one would not, by looking for Barr bodies be able to distinguish, if one saw no Barr body, an X–(XO or 45X) chromosome complement from XY; nor could one distinguish, if one Barr body were to be seen, a normal 46,XX complement from the abnormal complement XXY. The test in the newborn with ambiguous genitalia has these limitations and, in any event, the designation of gender is, as we shall see, more appropriately made on the appearance and potential functioning of the genitalia than upon the chromosome complement. One can get, however, more information concerning the sex chromosome complement in another way.

The Fluorescent Y: the Y body

Until about 15 years ago, it was impossible to visualise the Y chromosome complement quickly, easily and cheaply in non-dividing cells. There was nothing comparable to the Barr body and the n-1 rule. Now, since the work of Caspersson of the Karolinska Institute in Stockholm on fluorescent stains, it is possible to see the Y chromosome, and each and every Y chromosome in non-dividing cells.

Although in non-dividing, interphase cells the chromosomes are extended and indistinguishable, part at least of the Y chromosome remains compact enough that it can be seen as a discrete body if suitably stained with a fluorescent dye.

Buccal mucosal cells can be used but there is often so much fluorescent-staining debris that interpretation is difficult. Easier to interpret is a simple blood film stained, not with the stains used for differential white cell counts but with the acridine stains, either quinacrine mustard or quinacrine hydrochloride. With ultraviolet light illumination of the microscope the fluorescent Y chromosome can be seen to glow in the nucleus as a bright dot (Fig. 18). For the fluorescent Y the n-1 rule does not hold good. Each and every Y can be seen. Two F-bodies signifies two Y chromosomes (Fig. 19).

Using Barr bodies and F-bodies one can determine, almost in minutes, the X and Y chromosome complements. The two together may be of much more value than either alone in determining the genetic sex in a newborn of dubious phenotype.

The Autosomes

As long ago as 1912 Von Winiwarter claimed to have observed that humans have 47 chromosomes. He was nearly right. In 1923 Painter concluded that there were 48. He, too, was nearly right.

In 1956, Tijo and Levan in Sweden used tissue cultures of fetal lungs

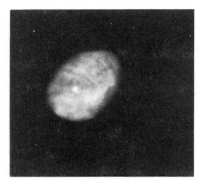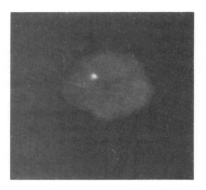

Fig. 18. The F body or Y body. The Y chromosome can be seen in an interphase cell—buccal mucosal cell on the left and a blood cell on the right.

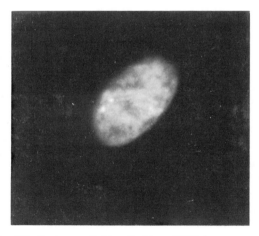

Fig. 19. Two Y bodies in one cell; this patient has a XYY chromosome complement.

to allow a clear look at dividing cells and to count chromosomes. They showed, and others soon confirmed, that the true number was 46.

Very soon methods were developed to study other and more mature tissues but, whatever tissue is used, the cells must be growing and dividing in tissue culture for it is only then, with one or two exceptions, that the chromosomes can be seen. Tissues that are naturally very rapidly proliferating and dividing can be used directly for chromosome study. Bone marrow is sometimes used when an urgent cytogenetic diagnosis may be wanted; to ascertain whether a newborn with duodenal atresia has Down's syndrome might be an example. But the quality of cells for cytogenetic diagnosis is much less good than when an *in vitro* culture is made. As we will see when we come to prenatal diagnosis, cells of the cytotrophoblast from the developing placental site are so rapidly dividing that the chromosomes can be seen directly and without culture.

Not very many tissues grow quickly or very well in tissue culture. Fibroblasts from skin and fascia take several weeks to grow adequately. Cells from amniotic fluid obtained by mid-trimester amniocentesis take about 3 weeks. Fortunately blood lymphocytes grow well and quickly. A chromosome study of cultured blood cells can be completed in about 3 days. Blood cells are used unless there is some good reason otherwise.

About 10 ml (but a much smaller quantity can be used in special circumstances) of venous blood is collected with a syringe or vacuum tube with sodium heparin. After very light centrifuging the white cells are separated off and added to a tissue culture medium to which is added an extract of red navy beans, phytohaemagglutinin (PHA), that stimulates growth of the lymphocytes. The culture contains antibiotics and antifungal agents to inhibit contaminant growth and, except for a special purpose, it contains folic acid. This tissue culture is set up with great care to ensure sterility and then is cultured for about 3 days.

The cells multiply by the same process of division by which damaged tissue is repaired and by which the fetus grows—mitotic division. Each cell, before, during and after division contains both members of each of the 23 pairs of homologous chromosomes, 46 in all (Fig. 20).

It is at metaphase (Fig. 20c and d) that the chromosomes are best seen. To prevent further progress to anaphase, colchicine is added to destroy the spindle fibres, to cause arrest of cell division at the stage of metaphase.

Next the cells are exposed to a hypotonic solution to swell them by osmosis. The chromosomes become further separated and distinct. After fixation with acetic acid, a film is made on a slide and dried in air. Stains of several types can be used to emphasise particular features of the chromosomes. After staining the slide is examined under the microscope and the best cells, the best metaphase spreads, are selected to be photographed. From a photographic enlargement such as Fig. 21, the chromosomes are cut out with scissors and arranged according to conventions agreed upon by cytogeneticists in a stereotyped array, a karyotype. The steps taken from blood sample to karyotype are shown in Fig. 22.

Sometimes it may be an advantage to study the chromosomes at an earlier stage of division than metaphase, in prophase. Such a variation

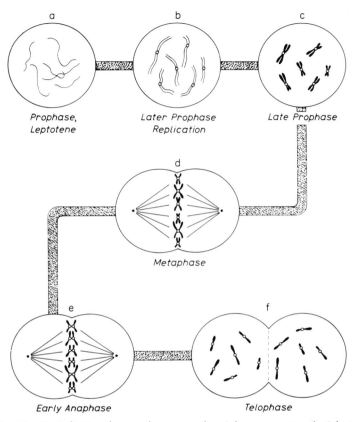

Fig. 20. *Mitotic division; there are the same number of chromosomes in each of the two daughter cells as there were in the original cell.*

has both advantages and disadvantages. The chromosomes are larger, longer, less condensed and their fine structure can be better seen. On the other hand, they are much more entangled, more overlapping, more like a mass of worms; their recognition as separate entities is much more difficult.

What one sees in Fig. 21 are chromosomes about to divide. At this stage each chromosome is composed of two chromatids, each with a long arm and a short arm, joined at a 'waist' or centromere. That they may appear X-shaped does not mean that they are X chromosomes. Indeed, if such a cell as shown in Fig. 21 were to be allowed to proceed to anaphase, each chromosome would be a single strand (Fig. 20f). It is the

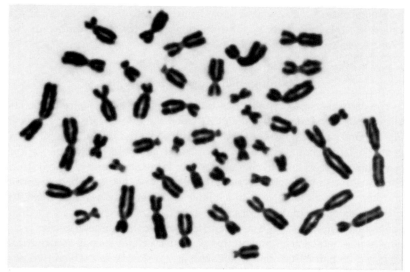

Fig. 21. *The chromosomes condensed and visible in a cell in metaphase of mitotic division in a tissue culture.*

method that we must use, and the stage at which we view them, that makes chromosomes appear as they are almost always illustrated.

Those chromosomes with the centromere, the 'waist', near to the middle are known as 'metacentrics', those with the centromere very near one end are the 'acrocentrics', those that are intermediate in shape are

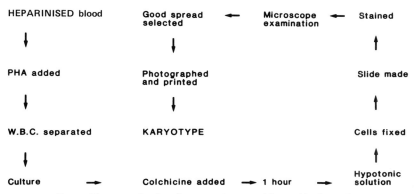

Fig. 22. *The steps whereby a karyotype is prepared from a blood sample.*

'submetacentric'. They vary in size from $1.5\,\mu$ to $7\,\mu$. The short arm is called 'p', the long arm 'q'. The short arm of the X chromosome, for example, is Xp; the long arm of the Y is Yq.

Staining, Banding and Karotype

In the early days of cytogenetics 'solid' or 'block' stains were used. The chromosomes were seen as in Fig. 21. Such staining has its limitations. In Fig. 21 one can, of course, easily see that there are large and small meta-centrics, large and small acrocentrics and submetacentrics of various sizes, but one can find, in that spread, six large acrocentrics (the bottom chromosome in the picture is one) that cannot be distinguished one from another. They all look very much alike. We can only say that there are groups: six large metacentrics, four large submetacentrics, 15 medium submetacentrics, six large acrocentrics, six small submetacentrics, four very small metacentrics and, in this particular cell, five small acrocentrics: 46 in all, 23 pairs. But one cannot with any certainty say which two of a group form an homologous pair. If one constructs a solid-stained karyo-type array one cannot say, within a group, which one truly matches with which (Fig. 23). Because of this limitation, such staining is now obsolete.

Almost all laboratories now use at least one of two 'banding' stains routinely, often both, and in special circumstances may use a variety of techniques. Most commonly used are 'Q banding' and 'Giemsa banding'. The former uses quinacrine as the dye and ultraviolet light for illumina-tion of the microscope. With this stain and this illumination, the chromo-somes glow green-yellow against a dark background (Fig. 24). Each chromosome, or rather each pair of homologues, shows a characteristic pattern of light and dark bands along its length (Fig. 25) which allow for definitive identification of the chromosome pair within a group. By a convention agreed by cytogeneticists at an International Congress, par-ticular banding patterns identify, let us say, chromosome pairs 13, 14 and 15 within the group of large acrocentrics otherwise only identifiable as group D (Fig. 23); similarly banding patterns can distinguish other members of each group (Fig. 26). Now every pair of chromosomes can be recognised and given a valid reference number.

Giemsa banding employs some method to make the chromosome take up the Giemsa stain. It may be prior treatment with trypsin: tryp-sin–Giemsa banding. Visible light is used for illumination and the appearance and photograph are rather different. The Giemsa bands, while characteristic for each chromosome, do not correspond exactly with the Q bands. Just what makes these banding patterns is not certain

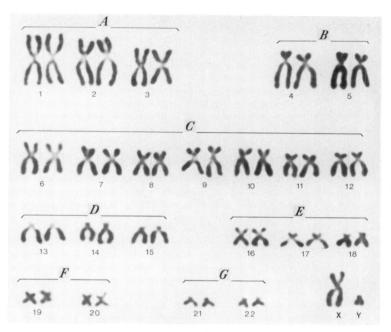

Fig. 23. An 'old-fashioned' block-stained karyotype; one cannot really distinguish the individual members of a group from one another.

but presumably they relate to variation in the degree of coiling or extension of the DNA molecule within the chromosome.

Which techniques are used in any individual laboratory depends as much on the experiences, preferences and, it sometimes seems, whims of the cytogeneticist. Q banding is quicker and easier than Giemsa banding which tends to be rather temperamental, and Q banding fades quickly under the microscope. Giemsa banding may show details not seen by Q banding and may demonstrate an abnormality that might otherwise have been missed. Each has its merits. Figs 27 and 28 illustrate karyotypes constructed by each of the two most usual stains.

Other methods may selectively stain the centromeres, may better demonstrate the rather indefinite ends of some chromosomes or may indicate exchanges of material between sister chromatids. These techniques have only occasional use.

Variants

While the chromosomes of one person look sufficiently similar to those

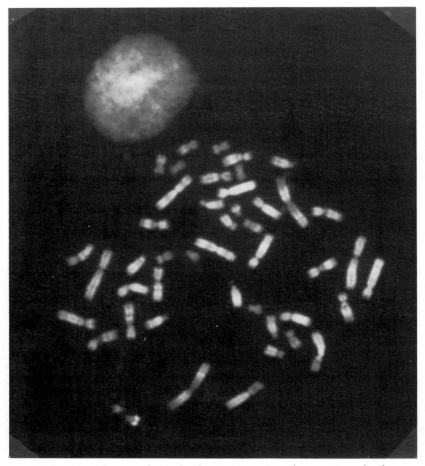

Fig. 24. A metaphase spread stained with quinacrine to give a fluorescent image by ultra-violet light.

of another that individual chromosomes can be given a certainly assigned group and number, there can be some variability in the same way as there is variation among human faces. The Y chromosome can be usually long, or short. It does not mean that the possessor of the big Y has exceptional sexual prowess. It is a variant of normal. A switch around, end-over-end, of a small segment of chromosome 3, including the centromere, is a common variant: pericentric inversion of chromosome 3. A pericentric inversion in chromosome 9 is sometimes seen.

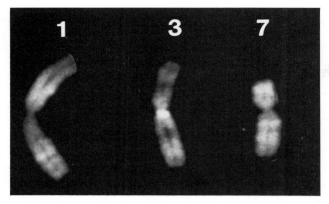

Fig. 25. *Chromosomes 1, 3 and 7 with distinctive banding patterns with fluorescent staining: Q banding.*

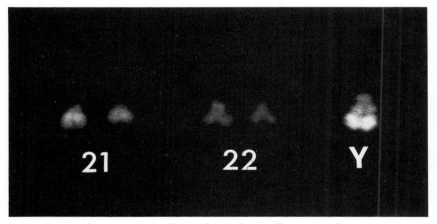

Fig. 26. *Fluorescent staining distinguishes the three small acrocentric chromosomes.*

These variants are inherited, contributed to the zygote, from parent to offspring in the chromosome complement of the parental gamete. One may be able to say of a pair of chromosomes which one was contributed by which parent. We will return to that later.

Band Mapping and Gene Mapping

Using the various forms of staining, and perhaps with not a little imagi-

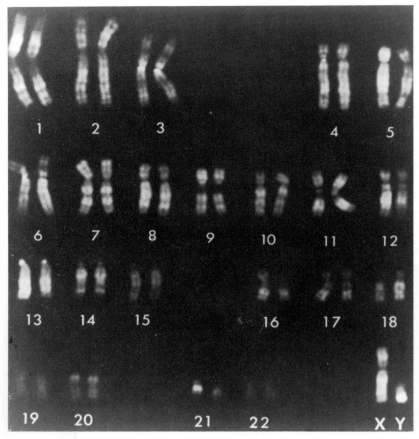

Fig. 27. A karyotype prepared after the chromosomes had been stained with quinacrine in metaphase.

nation, each chromosome can be assigned a detailed band pattern; each chromosome has its own peculiar spectrum of bands. For each chromosome (and for all the chromosomes) a band idiogram can be constructed. Figure 29 shows the idiograms of chromosomes 7 and X. Some cytogeneticists claim to be able to recognise some 400 bands in the whole karyotype. Some are less sanguine. Claims have been made to recognise over 2000 bands in the whole chromosome array as displayed by studying more extended, prophase, chromosomes—as opposed to the more usual metaphase chromosomes.

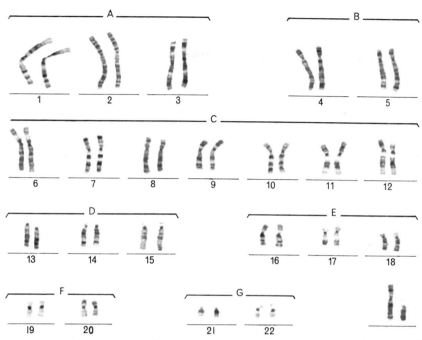

Fig. 28. A karyotype prepared after the chromosomes had been stained by Giemsa stain.

As we have said, p signifies the short arm of a chromosome, q the long arm. The arms are then divided, by cytogenetic convention, into regions: let us say, p1 and q3 of chromosome 7. Further to define a precise point on a chromosome, the band number within a region might be quoted in a text or report. It might be reported that there was an abnormality affecting chromosome 7 at 7q21. The 21 has, in this context, nothing to do with chromosome 21, the chromosome related to Down's syndrome. The 21 refers to the region and the band on the long arm of chromosome 7. One might encounter a reference to Xq28: the region and band near the tip of the long arm of X where are located the genes for G6PD, red-green colour blindness and haemophilia A.

Until about 15 years ago we knew very little about which genes are located on which chromosome. One could, to be sure, deduce from the pattern of transmission of an abnormal phenotype, Duchenne dystrophy or haemophilia for example, in a family pedigree that the gene must reside on the X chromosome: carried by females, transmitted to 50 per cent of the offspring, expressed in the (hemizygous) males and no father-

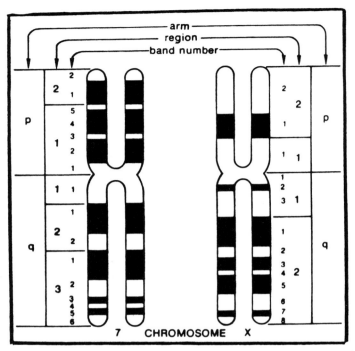

Fig. 29. *Chromosomes 7 and X indicating the long arms, q and the short arms p. Within the arms are regions and within the regions are bands designated by cytogenetic convention.*

to-son transmission. By about 15 years ago some 70 genes had in this way been assigned to the X chromosome; but no assignment had been made to any autosome.

In 1968 Donahue was investigating the hereditary transmission in his own family of a variant, a harmless peculiarity, of chromosome 1. He then recognised that this variant was inherited along with the Duffy blood group; there was, in 1q, a marker for the Fy gene. This was the first autosomal assignment. A variant in 6p in one family was found to be linked to the tissue types. The HLA loci and all their complexities were assigned to the short arm of chromosome 6.

Chromosome abnormalities, additions and losses of chromosome material, might be expected to show lessened or enhanced production of a protein or enzyme if the relevant gene is on the chromosome material that is lost or gained; and so it is. A triple, rather than the normal double, representation of chromosome 21, trisomy 21, in the chromosome complement is associated with increased activity of the enzyme superoxide desmutase (SOD). That gene can be assigned to chromosome 21.

It had been known from family studies that the genes for the ABO blood group system, the nail–patella syndrome, and for acid phosphatase synthesis were linked together. The three genes were inherited as a unit. But where, on what chromosome, did the cluster reside? Then Ferguson Smith in Glasgow found a person with double representation of the tip of the long arm of chromosome 9 who also showed increased synthesis of acid phosphatase. The gene for acid phosphatase and its closely linked associates could be assigned to the end, the termination, of the long arm of 9:9qter. Such fortuitous occurrences led to some autosomal assignments, but not many.

In 1968 a remarkable new technique was devised: somatic cell fusion, hybridisation. Metaphase cells from mice and men were mixed in culture with the Sendai virus. Some cells acquired mixed chromosome complements: mouse and man hybrid cells. As these hybrid cells grew in culture the less robust human chromosomes died out of the hybrid cells, one by one, randomly. As chromosomes dropped out, one could observe biochemical activities ceasing one by one. In this way, genes could be assigned to chromosomes. This somatic cell hybridisation technique caused an explosion of knowledge concerning gene locations.

As we have seen one can develop gene probes: traceable complementary sequences of DNA that can lock onto and thus signal the presence of a sequence or gene that is being sought. The somatic cell hybridisation technique can be used in conjunction with gene probes to 'sniff out' the genes in the one or two recognisable human chromosomes that are left in hybrid mouse–man cells when the others have been lost by attrition. But one does not now have to rely on hybrid cells to present one with individual chromosomes. There are other ways.

If one takes a metaphase cell on a microscope slide as in Fig. 24, the chromosomes can be recognised individually by the size, shape and banding pattern. If they are then treated chemically to open up their DNA to be single strands, we have made them into receptors for an appropriate complementary cDNA probe. If a gene probe, labelled with radioactivity, is applied to this treated metaphase cell, the probe will be taken up—as will the radioactivity—by the locus of the gene sequence on the chromosome. The locus of the looked-for gene can be detected by its radioactivity. One knows the chromosome by its morphology; we can make an assignment, gene to chromosome, by this probe-to-gene *in situ* hybridisation.

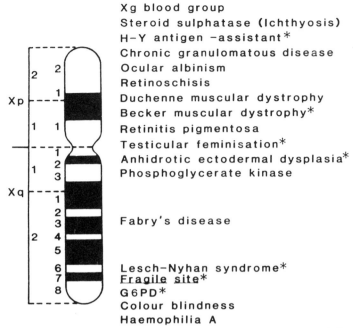

Fig. 30. Arms, regions, bands and some genes on the X chromosome; those marked * are referred to in the text.

Even this technique is not the ultimate. Chromosomes can be labelled in such a way that they can be sorted out from a mixture. One could, in theory at any rate, have a cupful of one kind of chromosome or another, 90 per cent pure! These then can be chopped into pieces with restriction enzymes and the gene sequences in the fragments studied with probes.

In these and other ways, the gene-map can be worked out. It is like a jigsaw puzzle. At first progress is very slow, but as more pieces fall into place, the pace of location of the pieces quickens and groups of pieces, clusters of linked genes, can be assigned to their proper place.

In just a very few years, maybe over a thousand genes have been assigned in the total gene map. Over 30 have been located on chromosome 1, over a hundred on the X chromosome; just a few locations are shown in Fig. 30. One has to read this week's scientific journals to keep abreast of the explosions of modern genetics. Actually, next week's would be better!

Stop press news! The gene for cystic fibrosis has now, November 1985, been located on the middle of the long arm of chromosome 7.

Recommended Further Reading

Gene Mapping

Shaws, T. B., Sakaguchi, A. Y. and Naylor, S. L. (1982). 'Mapping the human genome, cloned genes, DNA polymorphisms and inherited disease'. *Advances in Human Genetics, Vol. 12,* Chapter 5, Plenum Press, New York and London.

Steele, C. M. (1984). 'DNA in medicine. The tools, parts I and II'. *Lancet* **2**, 908–911 and 966–968.

White, R. L. (1984). 'DNA in medicine, human genetics'. *Lancet* **2**, 1257–1262.

CHAPTER 3

Collection of Specimens

Blood

There are few laboratory reports more frustrating than that which states 'Culture failed. Please repeat.' It can be difficult, sometimes impossible, or too late to repeat a sample for chromosome testing.

Blood samples for lymphocyte culture should be venous samples, taken with as strict aseptic precautions as for a blood bacterial culture. It is best taken into commercially available heparinised vacuum tubes. If these are not available, the laboratory to which the specimen will be sent should be consulted, as they may have special tubes available. It should be noted that 'regular' heparin, as used as an *in vivo* anticoagulant, is likely to contain preservatives that inhibit cell growth. Potassium heparin, as opposed to sodium heparin, also makes for failed cultures.

A single sample of 5–10 ml should be more than sufficient, but an extra tube may allow for accidents of breakage or contamination. The blood should be kept cool but must not be frozen. The sooner it is received in the laboratory the better, but the cells should be viable at 24 hours and culture might be successful as late as 48 hours. Courier service will be preferable to the mail, or the patient or relative—who have a personal interest in the result—could take the specimen to the laboratory.

The laboratory requisition to accompany the specimens should clearly indicate the nature of the problem to be elucidated by the laboratory. Do not simply request 'chromosomes please'! Cytogenetics laboratory technicians are an elite class. They are making most important value judgements as to whether an abnormality is present or not on the basis of what they see down the microscope or in the karyotype. They should know what they might be looking for: to confirm the presence of a whole extra chromosome is one thing, to recognize a tiny fragment of a chromosome missing, in excess, translocated to another chromosome or reversed in the banding pattern, an inversion, is quite another matter. A requisition stating 'baby with unusual dysmorphic syndrome; query some atypical chromosome abnormality' would ensure a more exacting scrutiny—and deservedly so—than one that stated 'Typical Down's syndrome; query confirmation and query trisomy or translocation'. Simple staining would suffice for the latter; special stains might be required for the former. If the problem is apparently X-linked mental retardation and the 'fragile X' chromosome is to be looked for, the requisition must state that fact. The culture medium for such a test must be quite different from that usually used. The laboratory must be given some indication what they might be seeking and why.

Skin Fibroblasts

Sometimes, because it may be suspected that more than one cell population each with different chromosome complements may be present in the body, it may be desirable to culture cells from some other tissue than blood. Fibroblasts from the deeper layers of the skin will grow, though very slowly.

A tiny fragment, no bigger than pinhead size, can be punched out. It only takes a second. The fragment can be put in sterile Hank's solution or, if not available, normal saline, kept cool and sent as quickly as possible to the laboratory. Skin samples travel better than blood. In Hank's solution they will keep for 3 or 4 days, and for 2 or 3 in saline. However, do not expect a report for 6 weeks!

Autopsy Specimens

It can be most important in genetic counselling to have an exact and aetiological diagnosis of an index case. Suppose a baby is stillborn or dies with multiple malformations or dysmorphic features, it may not be too late to study the chromosomes. The sooner after death the better, but even a week after death may not be too late if the body has been kept chilled. Two or three tiny fragments of skin and fascia should be placed in Hank's solution or saline and sent to the laboratory—with an appropriately informative requisition, of course. Vital information might be forthcoming in a few weeks.

Prenatal Diagnosis

The purpose of prenatal diagnosis—and almost always its happy result—is to reassure a couple that they do not have a pregnancy with some serious dreaded abnormality for which there may be exceptional risk. Prenatal diagnosis is only rarely a prelude to the possibility of termination of a pregnancy by an abortion. Prenatal diagnosis only rarely prevents the birth of a liveborn baby with an abnormality, but it does allow the birth of children who might otherwise never have been conceived by fearful parents. Certainly, a serious abnormality is sometimes found and the option of abortion has to be considered, an option repugnant to some. Nevertheless, my code of morals—and yours—are irrelevant. We must know, and we must tell our patients, what modern skills can offer; the acceptance or rejection of that offer are at the discretion of the parents alone.

It does not seem right to make a prior commitment to abortion a condition for prenatal diagnosis. Apart from the fact that a couple opposed to abortion are as entitled to know what is in store for them as are a couple who have made a decision that a handicapped child will not be born to them, a couple may change their minds. There is a world of difference between the question 'If you were to be told that your baby would have a severe handicapping birth defect, would you consider an abortion?' and the question 'You are going to have a baby with a severe handicapping birth defect. How do you feel about an abortion?' Any couple at exceptional risk of a pregnancy of an abnormal fetus must be offered prenatal diagnosis. Where there is dispute, the decision must rest with the mother.

The author does not consider the sex of the baby, male or female, one way or the other, to constitute an abnormality and, freedom of opinion notwithstanding, would not offer prenatal determination of sex for mere curiosity or to allow of selection by the parents of a child of the sex of their choice. Of course, a serious, untreatable X-linked disorder is another matter. Where a male, but not a female, child might have muscular dystrophy, fetal sex determination is medically relevant.

Amniocentesis

X-rays have very limited value in early prenatal diagnosis and fetoscopy, dramatic though the concept is, is a poor tool. It might, it is true, make possible a diagnosis of one of the several polydactyly syndromes or even of a cleft lip—if one were lucky—but only too often the view down the instrument is too limited, the amniotic fluid is too murky or one cannot manoeuvre the relevant fetal part into the field of vision. Moreover, fetoscopy carries about a 5 per cent risk of pregnancy loss. Fetal blood sampling by fetoscopy for diagnosis of the haemoglobinopathies or haemophilia has been made obsolete, or will soon become obsolete, by DNA sequence analysis of fetal cells obtained by amniocentesis or chorion villus sampling (CVS).

By far the most common reason for both prenatal diagnosis and amniocentesis is advanced maternal age. As we will see, certain chromosome abnormalities of the fetus become much more common as a mother attains what the prayer-book euphemistically terms, 'riper years'. Chromosome abnormalities of any kind can be recognised in metaphase cells grown from fetal cells obtained by amniotic tap.

Unfortunately, amniocentesis cannot be performed with a good hope of success at less than 15 weeks' gestation. Below that gestational age, there is insufficient fluid and 'dry taps', repeated taps, and a risk of loss of the pregnancy are more likely. On the other hand one does not wish to wait until 18 or 19 weeks. Amniotic fluid fetal cells (cells from the amnion, oropharynx, and urinary tract) take about 3 weeks to attain a sufficient number of metaphase cells for karyotype analysis. A tap done at 18 weeks will not give a result until 21 weeks, sometimes later. An abortion, if that becomes an option, is much more psychologically traumatic to the couple (and indeed to all concerned), more likely to be associated with complications and more likely to result in subsequent infertility the later in pregnancy that it is done. The time of choice for amniocentesis for chromosome analysis is between 15 and 16 weeks' gestation.

The position of the fetus and the location of a good pool of fluid is determined by real-time ultrasonography. The position of the placenta is determined. The best point of approach to a good collection of fluid but which avoids the placenta and the fetal head is marked on the mother's skin. As an outpatient procedure, without local anaesthesia, a fine (gauge 22) needle is inserted to a depth of from 4 to 6 cm depending on the position of the uterus, the location of the fluid and the obesity of the mother. Discomfort appears rather trivial though some mothers feel a sudden brief pain as the needle enters the amniotic cavity. Usually 10 to 15 ml of the pale yellow and slightly opalescent fluid are removed.

If there are twins or triplets, the amniotic sacs must be tapped separately for it is perfectly possible for one only of twins or triplets to have a chromosomal, or other, abnormality. And they can, of course, be of different sexes.

It is usual for mid-trimester amniocenteses to be done in medical centres near to a cytogenetics laboratory, but if such should not be the case, the amniotic fluid, untreated except to be kept cool but not frozen, can be sent by courier to the laboratory. Cells will grow, though less reliably, even after 3 or 4 days. In the laboratory, much the same procedure is followed as for blood cells (see Fig. 22). A karyotype is assembled, scrutinised for any abnormality and reported about 3 weeks after the procedure. In the course of the chromosome analysis, of course, the sex of the fetus will become known, XX or XY. Most couples wish to know what sex can be expected; a few like a surprise.

A study of chromosome morphology is not all that can be achieved by amniocentesis. The molecular structure of the chromosomes can be studied by DNA technology. It may be possible, as previously discussed, to acquire or build up a gene probe that will directly identify an abnormal gene sequence in the chromosome complement or, failing the existence of a direct probe, it may be possible to infer, with greater or less degree of certainty depending on the closeness of the linkage, the presence of an abnormal gene by probing, not for the gene itself, but for a marker RFLP. If there are sufficient cells, sufficient chromosomes, sufficient DNA, in the amniotic fluid, a DNA analysis can be done on DNA extracted from those cells, in a day or two, without culture. If not, the DNA can be augmented by culture of the cells and by their synthesis of DNA as the chromosomes replicate at cell division.

While not strictly relevant to this text, one should know that many biochemical activities, either directly or after culture, of amniotic fluid cells can be examined. Many (over 60) inborn errors of metabolism can be prenatally diagnosed in this way, though one suspects that many of these tests of biochemical cell phenotype will be replaced by DNA genotype determinations. Finally, one might know that the major 'open' neural tube defects, anencephaly and open spina bifida, can be diagnosed by excessive leakage through the surface defect of fetal alphafetoprotein into the amniotic fluid.

What are the risks of amniocentesis? In expert hands, at the right time, very small. Risk to the mother and risk of direct trauma to the fetus can be considered 'negligible'. The hazard is that the procedure might cause fetal death and loss of the pregnancy. Figures vary from centre to centre and depend to some degree on biases of case selection. It seems to this author that, in good hands and with a placenta out of the way, the risk of fetal loss can, as some studies have shown, also be described as 'negligible'. With a placenta so anterior that transfixion cannot be avoided, the risk of loss could be as high as 1 per cent. Many people quote an overall fetal loss rate for amniocentesis as 0.4 per cent or thereabouts.

Chorion Villus Sampling (CVS)

Because it must necessarily be done so late in pregnancy and because results are so slow forthcoming, well-tried and safe though it is, amniocentesis is far from ideal. We need something as safe, as comprehensive and as reliable but something that can be done earlier and with quicker results. Chorion villus sampling appears to offer these benefits. The technique, pioneered by Han An-Guo of the Tietung Iron and Steel Company Hospital, China, in the late 1960s and early 1970s, has been slow to reach the western world, via the USSR. It has only been applied outside those countries since about 1981, notably by the Italians.

Up to about 6 weeks' gestation, the embryo in its celomic sac is surrounded by chorion. From about that time onward an amniotic cavity develops and one part of the chorion proliferates and thickens to becomes, like a mass of seaweed, the chorion frondosum. This will become the localised, definitive placenta. The remainder of the chorion becomes thinner, flatter and a mere membrane, the chorion laeve.

This chorion frondosum is fetal tissue, cytotrophoblast, rapidly growing by exuberant cell division, invading the maternal decidua basalis. It is fetal tissue in its chromosome complement, in its genomic DNA and in many of the biochemical activities of its cells. It represents the fetus as amniotic fluid cells do also.

At about 8–10 weeks, best perhaps at 9 weeks, this chorion frondosum can be sampled without, so far as one can see, great hazard to the fetus. Techniques vary in matters of detail, but most operators use an approach through the vagina and cervix.

As an outpatient procedure, without anaesthesia or much discomfort, but with very high quality ultrasound guidance, a thin plastic (some people use malleable metal) catheter is passed through the cervix and into the chorion frondosum (Fig. 31). Gentle suction and slight movement of the catheter break off small fragments of chorionic villi often with some adherent maternal decidual cells. They look like tiny pieces of pale seaweed (Fig. 32), and together the fragments might weigh between 20 and 50 mg. That is enough; much can be done with this fetal tissue.

At the time of the procedure, the aspirated material is examined under a dissecting low-power microscope to ensure that it is indeed villous tissue. The blood clot and maternal tissue is dissected away. If chromosomes are the matter of relevance, they can be studied directly for this is such rapidly growing tissue that there are in it, without culture, many cells in mitosis and some in natural metaphase. The cells are freed from the supporting framework of connective tissue, fixed, stained, examined, photographed and the chromosomes counted. It sounds easy, but it is not quite so simple. The main problem is to find enough good and unbroken cells at the right stage of mitotic division. Some of the afficionados of CVS use computerised slide scanners to find their best cells for

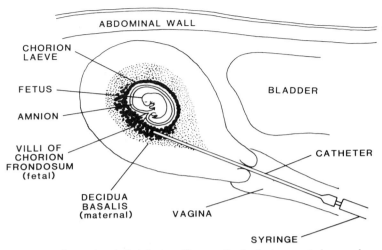

Fig. 31. *The usual method of chorion villus sampling by the transcervical approach.*

them, but with a good sample it is possible to be looking at chromosomes within the hour and to give a report in just a few hours, not weeks as for amniocentesis. If a direct preparation of natural mitoses is not satisfactory, or to confirm results, a short-term culture of cells can be set up. There is a snag: some contaminant maternal cells can grow and cause errors and confusions. The direct preparation does not give this problem for maternal tissue is not rapidly dividing. There are no maternal cells in metaphase. Most workers are trying to get better direct preparations.

How accurate is CVS for chromosomal study? We are not sure as yet. It seems agreed that the pictures are less clear and interpretation more difficult with CVS than with amniocentesis. That problem may be overcome. There is also some anxiety that CVS may on rare occasions show an abnormal chromosome complement in a fetus whose body cells have no abnormality. There is also anxiety that normal fetuses could be aborted on the basis of a 'false positive' CVS. The answers are not all complete.

Cells from CVS can be used in other ways. DNA gene analysis has been applied very successfully using either specific gene probes or marker RLPLs to make an early genotypic prenatal diagnosis. The cells can be used also for biochemical studies and the detection of inborn errors of fetal metabolism. Indeed CVS cells may be better than cells from amniocentesis. They may exhibit a phenotype, a biochemical activ-

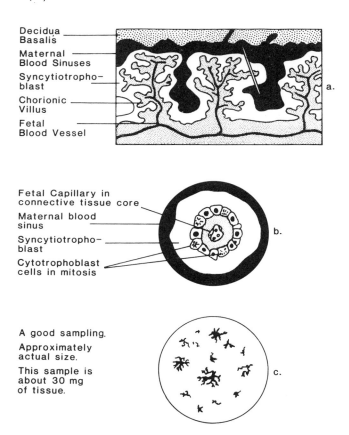

Fig. 32. Chorion villus sampling. The line in (a) indicates a villus seen in cross-section in (b); (c) shows the approximate real size of the fragments of a good villus sample.

ity, that has been 'switched off' in amniotic cells; but, again, there can be the problem of maternal cell contamination.

What of the risks? What of the disadvantages? We just do not know. Why not? Let me explain. Risks due to a procedure must be assessed in relation to the 'background' or natural risk. What is the risk of fetal loss if the procedure had not been done? The loss attributable to the procedure is a risk over and above this natural fetal loss. Woman lose pregnancies frequently. Miscarriages happen and did happen before CVS was invented. The period of gestation at which spontaneous miscarriages most commonly occur is about the time, or shortly thereafter, when CVS is done. There is another difficulty. One would have no reason to be doing a CVS if the fetus had been shown to be dead, or destined shortly to be spontaneously miscarried. The 'background' for comparison must be the risk of spontaneous abortion, without CVS, of a fetus known to be alive in the period following the time at which CVS usually is done. But there is yet another problem. If CVS reveals an abnormality, the pregnancy will probably be terminated by intervention before it would have had the opportunity of being spontaneously miscarried. Therapeutic abortion will reduce the miscarriage rate in those women who have had CVS. But even that is not the end of the difficulties of determining fetal loss due to CVS. Spontaneous miscarriage is very much related to maternal age. What, then, is the background natural fetal loss of fetuses alive at a time when CVS is done in groups stratified by maternal age, taking no account of the possibly important variables of parity and socioeconomic class!

Figures are beginning to emerge, and it looks as though the background rate of fetal loss overall is about 2.5 per cent, with much higher rates of spontaneous loss in women over 40 years of age.

What, then, is the additional loss truly attributable to CVS? Even that figure is hard to come by. If the added risk were to be quite small, comparable, let us say, to the risk due to amniocentesis, huge numbers of CVS cases and controls would need to be compared to achieve statistical validity. Only if the added risk is great will that added risk soon become apparent.

What can one say at this point (March 1986)? It looks like the risk of fetal loss directly due to CVS is very small, perhaps no more than that ascribed to amniocentesis: let us say 0.4 per cent. But figures are not all that matters. Who can judge the mental anguish that a woman might feel who having had CVS has had, of her own decision, a therapeutic abortion of a fetus shown by the CVS to have an abnormality, as compared with the resigned disappointment of a woman who has not had CVS but has lost a comparably abnormal fetus by a spontaneous abortion over which she had no control.

These things are not easy, but CVS, one thinks, is here to stay. The great advantages of an early procedure and a quick result will outweigh the uncertainties and disadvantages unless these are very great. Let us hope public clamour does not demand the widespread use of CVS before some of these uncertainties have been resolved by carefully controlled and statistically reputable studies.

If I have dwelt at inordinate length on CVS it is because it is a procedure that has captured public attention while the problems are still poorly understood by the medical profession.

Circulating Cytotrophoblast Cells

This may come to nothing, but as a non-invasive prenatal diagnostic procedure it might offer great things. It is possible to recover fetal cytotrophoblast cells from the maternal circulation: cytotrophoblast emboli, if you will. If these fetal cells could be grown in culture—and there seems to be no reason to suppose otherwise—we might have a maternal blood test for fetal abnormalities, chromosomal and otherwise. We shall see.

Alphafetoprotein and Chromosome Disorders

In 1983, a woman asked her physician if a low alphafetoprotein level found in her serum (MSAFP) in her pregnancy could be related to the birth of her baby with trisomy of chromosome 18. The matter was taken up. It was found that the MSAFP was below the median values for unaffected pregnancies in 43 of 53 pregnancies of babies with chromosome anomalies of various kinds. Another study of the relationship, if any, of low MSAFP values to Down's syndrome (trisomy 21) showed that the mean MSAFP value for that chromosome anomaly was at the 25th percentile for unaffected pregnancies. The difference between MSAFP in normal and Down's syndrome pregnancies was highly significant. Further investigation appears to indicate that the MSAFP reflects the level of amniotic fluid AFP (AAFP). It is hypothesised that immaturity or malfunction of the fetal liver in certain disorders, including chromosome abnormalities, may lead to defective synthesis of AFP in the fetus and thus to low AAFP and to low MSAFP. It is moreover suggested that high MSAFP values make it very unlikely that the pregnancy is of a fetus with a chromosome abnormality.

Further research is needed, but it may be that MSAFP testing in pregnancy may be at least as useful in warning of the possibility of a chromosome abnormality as it is in warning of a neural tube defect. This is an exciting new possibility.

Recommended Further Reading

If there are no suggestions of a text for further readings to supplement this chapter, it is because there is no up-to-date book that covers all these matters. The foregoing views are a personal distillate of many scattered journal articles and conference proceedings. I can only refer the reader to the journal, *Prenatal Diagnosis,* and to the *Lancet* over the last 3 years—not neglecting the Letters to the Editor, a most valuable source of information and controversy. A seminal paper on chorion villus sampling is that of Simoni, G., Brambati, B. *et al.* (1983). 'Efficient direct chromosome analyses and enzyme determinations from chorionic villi samples'. *Human Genetics* **63**, 349–357.

CHAPTER 4

Normal Cell Division

When a species reproduces sexually a new individual exists from the moment when sperm and ovum unite at conception. These gametes unite to form the zygote. Since the zygote is the first of the myriad of cells that by repeated divisions make up the confederation of the body, genetic instructions must be conveyed to each gamete, and thus to the zygote, so that those repeated divisions, differentiations and proliferations take place as they should and so that functions can be taken up and maintained. The genetic codes of both parents must be incorporated in this first cell and passed to the generations to which it will give rise.

As we have noted, every body cell contains 46 chromosomes, 23 pairs. For the conception of a new individual with 46 chromosomes the formation of gametes, gametogenesis, must require halving of the chromosome complements in the parents' gonadal cells. This is a reduction cell division: meiosis at gametogenesis.

Meiosis: How it Works

Before considering the details of meiosis one should make some general observations. Let us first take male gametogenesis.

The sperm-producing cells start into meiotic activity at puberty. Each primary spermatocyte divides, by the first meiotic division, into two secondary spermatocytes. These each in their turn divide into two sperms by a second meiotic division: one primary spermatocyte, four sperms. Sperms are made continuously in bountiful supply. The two meiotic divisions and the period in between take about 2 months. There is no long delay in the manufacture of sperms. The turnover is rapid. In the female, it is quite otherwise.

A female is endowed with her full complement of germ cells when she herself is still a fetus and it is even in the fetus that female gametogenesis begins. The oocytes start the first stage of gametogenesis, the prophase of meiosis, even before birth. From then on they remain quiescent, biding their time, waiting long years for their debut at ovulation and their fulfilment, perchance, at fertilisation.

57

Before each ovulation one oocyte in prophase (occasionally more than one as evidenced by fraternal twins or triplets) is selected to awaken from its long rest of 15 or maybe 50 years. Meiotic division is resumed. As this first meiotic division now proceeds, one cell takes all the substance of the primary oocyte unto itself. The other, mere chromosome material, withers away as a 'polar body'.

At ovulation this well-endowed oocyte proceeds to the second meiotic division, a division that is not completed until it encounters a sperm. Then, with a whole new life before it, it completes division into another polar body and into the fertilised ovum, the zygote—one ovum from each primary oocyte.

To recapitulate, male gametogenesis is a brisk and bountiful affair. In the female it is much more leisurely. The prophase of the first meiotic division lasts for years. There is time for things to go wrong.

Meiosis in Detail

In the early stages of the first meiotic division in both males and females (in adult life or fetal life respectively) the chromosomes become condensed in leptotene (Fig. 33a). At the next stage, zygotene (Fig. 33b) the two representatives of each pair of chromosomes (which were each derived from the gamete of each parent of this individual who is making gametes in his/her turn) come to lie side by side.

Having paired, each chromosome builds onto itself a replica. Each chromosome now is two chromatids joined at a centromere (Fig. 33c). Though double, it is still a single chromosome.

Later in pachytene the chromosomes become more condensed while still side by side. We have indicated by drawing in open line or by blocking in that one chromosome came from this person's father, the other from the mother: next comes the most important stage, the stage that indeed has led to continual variation within a species, to infinitely variable combinations, to the uniqueness of the individual, to aptitude or ineptness in adapting to the environment in struggles for success, and to the potentiality for evolutionary progress as those with the best genetic advantages are the most favoured in reproductive performance. This is the phase of diplotene and diakinesis: the exchange of genetic material between the homologous chromosomes and their homologous gene loci. While concerned with the same matters of genetic business, the genes at these homologous loci may have non-identical alleles, alternative instructions in matters of detail.

At this stage (Fig. 33e and Fig. 34) the chromatids in conjunction

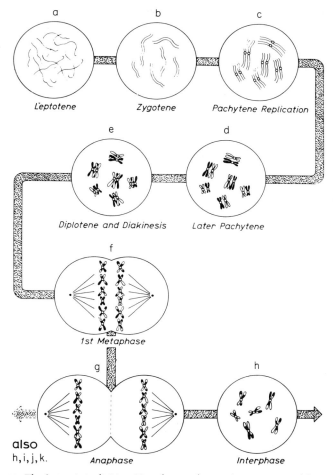

Fig. 33. *The first meiotic division. Note the interchange of genetic material between the maternal and paternal chromosomes in (e); note also that at the end of this first meiotic division (h) each chromosome consists of two chromatids joined at the centromere.*

break along their intertwined lengths at chiasmata and bits and pieces are exchanged at crossing-over. New chromosomes, new recombinants, are formed, uniquely made by the exchanging at random of genetic material that came from this person's parents. Some may confer genetic and evolutionary advantage.

Now we can see the relevance of the position of genes on a chromosome. The closer they are together, the closer the gene linkage, the less

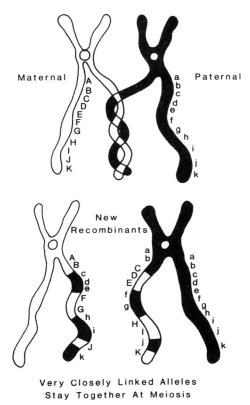

Maternal   Paternal

New
Recombinants

**Very Closely Linked Alleles
Stay Together At Meiosis**

Fig. 34. Crossing-over and interchange of genetic material at synapsis of diakinesis; genes located close together are unlikely to become separated at this crossing-over. This close, or loose, linkage is a most important concept in modern genetics.

likely is it that a break point will come between them; the more likely is it that they will move as one. The more likely is it, to return to our earlier analogy, that Tom and Dick will be together in the forest; the more likely that the detection of one can be used to infer the presence of the other. This is why it is so important that the detectable RFLP known as G8 is very near to the, as yet, undetectable gene for Huntington's chorea. DNA gene probes have made these matters of linkage and crossing-over of paramount importance in modern genetic practice.

At the next stage, first metaphase (Fig. 33f), the spindle fibres pull the new chromosomes into each new cell as anaphase completes the first

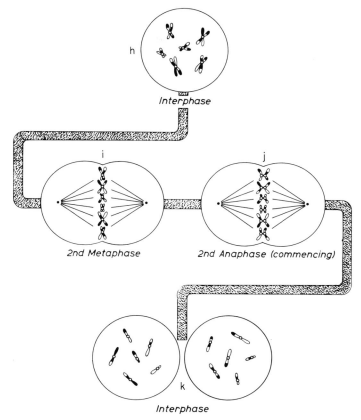

Fig. 35. *The second meiotic division; at anaphase the centromeres divide and each chromatid becomes a chromosome in its own right.*

meiotic division (in females at the time of ovulation). Each cell now has half of the chromosome complement, half of each pair, the haploid rather than the diploid number. There is more to come.

After a period of interphase, meiosis starts again (Fig. 35). Again the chromosomes line up in single file (Fig. 35i). They have no partners, each is a partner unto itself. Each makes its own partner by the splitting of the centromeres into two (Fig. 35j). At the second anaphase these now separate chromosomes, chromosomes in their own right by virtue of owning their own centromere, migrate into separate cells: the new finished products, the gametes.

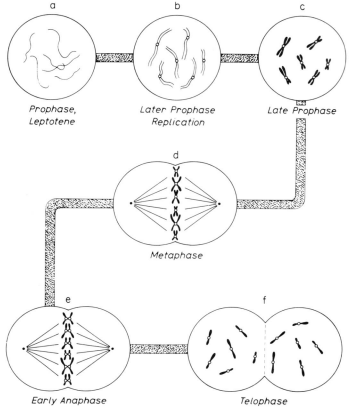

Fig. 36. Mitosis recapitulated from Fig. 20 to illustrate that the second meiotic division is comparable to mitotic division; in each the centromeres divide and each chromatid becomes a new chromosome.

Mitosis

Mitosis is that form of cell division by which the zygote grows into an embryo, an embryo into a fetus and a fetus into a baby and child. It is also the process by which cells divide in tissue culture in our chromosome preparations. Mitosis does not entirely cease with the completion of the body; in some tissues it is a continuing process as continuing repair restores continuing delapidation. Cells in the mucosa of the intestinal tract have a lifespan of just a very few days. They age, die and are replaced by active mitoses. Indeed, normally this mitotic activity can be observed in the crypts of Lieberkuhn in the duodenum. The chromosomes in metaphase can be seen, somewhat indistinctly it is true, in an ordinary histologic section. Some cells in some organs retain an option on mitosis. In liver damage from disease or injury, in skin and connective tissue damage, cells that have retained this option can repair an injury. Some cells have no such option. Muscle cells and neurones cannot multiply once their final number has been established. Sometimes cells assume a mitotic activity for which there is no good purpose; tumours, benign or malignant, are the result.

Errors of mitotic division play a lesser part than meiotic errors in causing the chromosome disorders, but mistakes at mitosis do have, as we shall see, a particular relevance.

The process of mitosis is comparable to the second meiotic division. The chromosomes contract in early prophase, replicate in later prophase, appear as we see them in our chromosome preparations and, at anaphase, separate as new separate chromosomes as each centromere splits along the long axis of the chromosome. One centromere, one chromosome, is the rule (Fig. 36). Each chromosome is formed by the long arm and the short arm of each chromatid with the centromere at some point along the length: metacentric, submetacentric and acrocentric chromosomes.

In mitotic division (and in the second meiotic division, for that matter) there is no reduction in the chromosome number. The number remains diploid. The number of chromosomes is the same in Fig. 36f as in Fig. 36a. The total chromosome complement, the full genome, is passed from generation to generation of dividing cells.

CHAPTER 5

Abnormal Chromosome Complements

Chromosome Abnormalities: Definition

Before embarking on descriptions and explanations of abnormal chromosome complements, some definitions of terminology are in order. *Euploidy* means a normal number of chromosomes or a multiple of the normal number: 23 pairs. A *triploid* complement, while euploid, would have three representatives of each pair: 69 in all. *Tetraploidy* would indicate 92 chromosomes. *Pentaploidy* and *hexaploidy*, five or six complete sets, have been observed. *Aneuploidy* describes a departure from the normal number, 47 chromosomes for example, or 45. The *diploid number* is 46, two representatives of each of the 23 pairs. The *haploid* number is when, after meiotic reduction division at gametogenesis, one representative of each pair is present in the gamete.

Trisomy refers to triple representation of a particular chromosome. Trisomy 21 indicates an extra chromosome 21, three of a kind, 47 in all. *Monosomy* is the opposite, one of a kind, 45 in all: 45X or 45,XO as it is sometimes written. *Deletion* means the whole or a part of a chromosome missing; *addition* means an excess of part or of the whole.

A *chromosome abnormality*, therefore, can be defined as a situation where there is an abnormal number of chromosomes without there necessarily being any detectable phenotypic abnormality in the possessor. All monosomies, and they are very, very few in liveborn humans, are associated with significant defect, but certain trisomies in some individuals are not; yet we regard all trisomies as 'chromosome abnormalities' or 'chromosome anomalies'.

A chromosome abnormality is also any departure from the usual size, shape, or banding patterns of a chromosome if, but only if, it is associated with a phenotypic abnormality in the possessor or the offspring of the possessor. We have to make that proviso to exclude variants. A pericentric inversion of chromosome 3, while a peculiarity, is not an abnormality. A large Y chromosome likewise is a variant of normality.

The abnormalities that for the most part concern us are abnormalities in one or other gamete, transmitted to the zygote and perpetuated there-

64

after in subsequent generations at mitotic division and cell multiplication. The error in the gamete can occasionally come about because the maker of these gametes has himself/herself a chromosome anomaly in all body cells, including the germ cells, and that anomaly passes on to the gametes; but the error can, and more usually does, come about by some error at meiotic division, at gametogenesis in a perfectly normal person. When one considers the breaks, crossing-over and rearrangements that normally happen at meiosis it is not difficult to see how sometimes bits and pieces can go missing or become misplaced; it is not hard to see how whole chromosomes might migrate incorrectly at anaphase.

Inversion

It can happen that a person, male or female, may have a segment of a chromosome switched around, end-over-end. We have noted that harmless pericentric (involving the centromere) inversions of chromosomes 3 and 9 are found in normal people as variants. They are harmless because the switched segments are genetically silent. Neither the possessor nor offspring are affected.

But suppose a switch-around involves a genetically active section of chromosome; what then? No harm to the possessor; they have all their chromosome material, all their genes. They are, themselves, quite normal. It is in the offspring that problems can arise. Meiosis, chiasmata formation, crossing-over and the creation of new recombinant homologous chromosomes present problems.

Figures 33e, 34, 35 and 37a show normal interchange of homologous segments and their genes as maternally derived and paternally derived chromosomes lie together and exchange piece for piece. In Fig. 37b there is a problem. How can there be exchanges if homologous segments do not line up? The answer is 'with difficulty'. Loops must form, 'inversion loops' and, with the formation of these loops, a single cross-over can cause genetic chaos in the gametes. Very abnormal recombinant chromosomes, with very abnormal genetic compositions, can be incorporated into the gamete. An entirely normal person, carrying a genetically significant inversion can have abnormal (or, if they are lucky, normal) children, perhaps time after time. They have a 'balanced' inversion. A child may (Fig. 37b.3) acquire a balanced inversion like the parent; or it may be 'unbalanced' (Fig. 37b.2,4). Sometimes prenatal diagnosis may require one to distinguish, from the shapes and sizes of the chromosomes, the one from the other. It may not be easy.

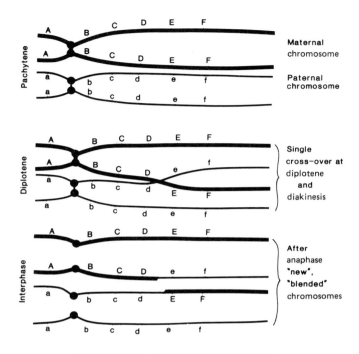

SINGLE CROSS-OVER AT NORMAL MEIOSIS

Fig. 37(a). Single cross-over at the first meiotic division. Capital letters indicate maternal gene sequences, small letters the homologous paternal sequences. This figure is a restatement of Fig. 34, but with a single cross-over.

Deletion

Sometimes it can happen in the shuffle of chromosome segments at meiosis that a piece can go missing at crossing-over. It does not become reattached to the body of the chromosome. If it is unattached, if it has no centromere, if it is an 'acentric' fragment, it and its genes are lost to the chromosome complement (Fig. 38a). The parent is normal. The offspring has a *de novo* (in it only) deletion and is likely to be quite abnormal.

Sometimes fragments can break off from both ends, and both become lost as acentric fragments. The bare ends of the chromosome can unite to form a ring: a *ring chromosome* (Fig. 38b). Clinical defects result not so much from the formation of the ring as from the deletions that have made the ring possible. Denuded ends can stick together; complete ends do not.

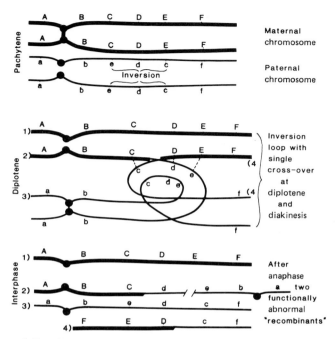

1) **Normal, maternally derived**
2) **Dicentric (two Centromeres); will probably break into two chromosomes**
3) **Inversion in paternally derived chromosome. Functionally normal**
4) **Acentric fragment. Will probably get "lost" at anaphase because no centromere for spindle to attach to.**

PARACENTRIC INVERSION CAUSING PROBLEMS AT MEIOSIS

Fig. 37(b). When there has been an inversion of a segment of a chromosome the chromosomes must undergo contortions to bring the homologous segments opposite one another. Crossing-over of these inversion loops can give rise to abnormal chromosomes.

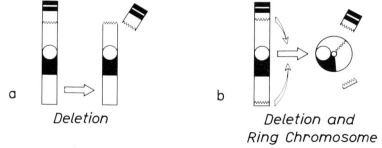

Fig. 38. In (a) there has been a simple deletion, simple loss of genetic material; in (b) there have been two break-points and two 'sticky-ends' so that a ring has formed. It is the deletions that are important, not the ring formation.

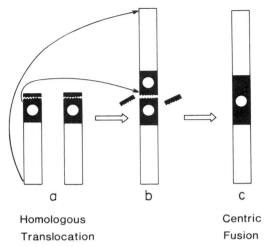

a   b   c

Homologous   Centric

Translocation   Fusion

Fig. 39. Two homologous chromosomes have each suffered breaks and have joined together as an homologous translocation. The centromeres have then fused together so that one now has a single chromosome consisting of the genetic material of almost two whole chromosomes.

Translocation

Homologous

It can happen, though rarely, that a person, male or female, can have two homologous chromosomes stuck together, both of the pair translocated. Usually this happens with the acrocentric chromosomes, not uncommonly a fusion of both chromosomes 13: a 13/13 translocation (Fig. 39a and b). Usually the centromeres become fused, 'centric fusion'. Such a translocation is sometimes called a Robertsonian translocation.

This translocation does no harm to the possessor. The tiny acentric fragments lost have little or no gene activities. Gametogenesis presents problems (Fig. 40). Both representatives, translocated as they are, must go into one gamete or the other. If fertilised or fertilising, the zygote will have three representatives of the pair (though only 46 chromosomes as they could be counted) or one (and 45 as they are counted). A zygote would have either a translocation trisomy or a translocation monosomy. A person with a homologous translocation, while normal themselves, cannot have normal children. There are such cases.

Non-homologous, Whole Chromosomes

What if a whole chromosome is translocated to another in the comple-

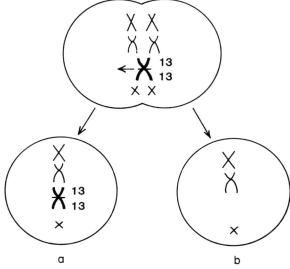

Fig. 40. Homologous translocation of chromosome 13, 13/13 translocation. At gameto-genesis both representatives of chromosome 13 move into one gamete leaving the other gamete with no representative. Zygotes will be either trisomic for chromosome 13 or they will be monosomic.

ment: let us say, as a fairly common example, one chromosome 21 translocated to chromosome 14. The possessor of such a balanced translocation is normal. They have all their genes; they are only rearranged. But that rearrangement can give great problems (Fig. 41). Gamete A1 has two representatives of chromosome 21. Fertilisation will add another. There will be a translocation trisomy of trisomy 21. Gamete A2 will have no representation of 21. The zygote have a translocation monosomy. Gamete B1 will have a single representative of 21 but it will be translocated. The zygote will have a balanced translocation like the parent. It will be normal but will, in its turn, have problems with its offspring. Gamete B2 will be normal in chromosome number, normal in distribution. In theory, any offspring has an even chance of being abnormal in phenotype because of either translocation trisomy or monosomy. As we shall see later, it does not quite work out that way. So much the better.

Non-homologous, Fragments
A rather common problem is that presented by the person who has part

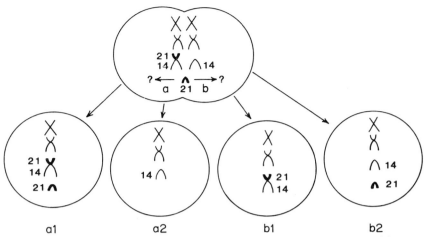

Fig. 41. *Gametogenesis in a balanced 14/21 translocation carrier. Four gamete possibilities exist; thus there will be four zygote possibilities: one trisomy, one monosomy, one balanced translocation carrier and one with a normal chromosome complement.*

of one chromosome translocated to another: let us say, part of the long arm of 13 onto the short arm of 3. That person is normal; they have a balanced translocation. Again, gametogenesis presents problems (Fig. 42). Gamete A has the normal 13 with part of the long arm of the other 13 carried by the short arm of chromosome 3, 3p +. Gamete B has 13 with the deleted long arm but with chromosome 3 carrying that segment on the short arm. Gamete C has the deleted 13 with the normal 3 chromosome. Gamete D is normal in all respects.

A zygote derived from A will have triple representation of part of the long arm of 13. It will be abnormal. Zygote B will have a balanced translocation and, like the parent, B will have a normal phenotype. A zygote derived from C will have only one representation of part of 13q. It will surely be abnormal. A zygote from D will be normal in all respects.

Such cases are not rare. This 3p + was in one of my patients. We will return to him later.

Isochromosomes

As we have seen in Fig. 35 a chromosome in metaphase consists of two chromatids joined at the centromere. At anaphase the centromere splits longitudinally (Fig. 43a). There can be errors at this division. The split can be horizontal (Fig. 43b). Each new chromosome will consist of either

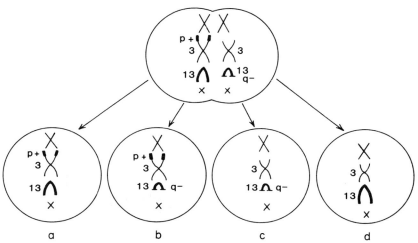

Fig. 42. *Translocation of part of a chromosome, in this case translocation of part of the long arm of chromosome 13 to the short arm of chromosome 3. Again four zygote possibilities exist: one will be trisomic for part of the long arm of 13, one will have a balanced translocation like the parent, one will be monosomic for part of the long arm of 13 and one will be entirely normal.*

both long arms joined at their own centromere, or both short arms with a centromere. Two abnormal chromosomes are formed: an isochromosome of both the long arms and the short arms.

A gamete thus can have double representation of the long arms, but no representation of the short arms; or it can have double representation of the short arms and no representation of the long arms. Fertilisation by a normal gamete will give a zygote with, in effect, trisomy of the long arms and monosomy of the short arm, or monosomy of the long arm but trisomy of the short arm. Such unbalanced chromosome complements will cause clinical abnormalities.

Anaphase Lag

It can happen that at anaphase a spindle fibre does not 'pull' a chromosome correctly into its cell. It can lag behind its fellows and drop out (Fig. 44). Fertilisation of a gamete that has lost a chromosome in this way will result in a zygote monosomic for that chromosome. Monosomy for a whole chromosome is so catastrophic to the zygote that, with rare exceptions, liveborn babies do not result.

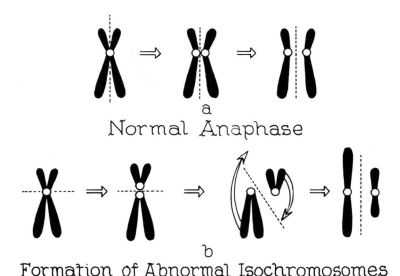

Normal Anaphase
a

Formation of Abnormal Isochromosomes
b

Fig. 43. Isochromosome formation: the centromere has divided horizontally rather than vertically so that the new chromosomes are made up of both of the long arms and both of the short arms of the dividing chromosome.

Non-disjunction

Primary

Note, first, the spelling of the word; it is not, as is commonly written, 'dysjunction'. The term disjunction refers to migration of the chromosomes at anaphase, each into its own new cell.

Sometimes that process goes wrong and both members of a pair of chromosomes migrate together into one gamete, and none into the other (Fig. 45). Fertilisation will result in a trisomic zygote on the one hand, monosomic on the other. As we shall see, this is the most common of all the chromosome anomalies.

Secondary

Suppose all the body cells should be trisomic, including the cells of the gonad, what might be expected at meiosis? Three cannot evenly be divided by two. Two members of the trisomic chromosome move into one gamete, one into the other (Fig. 46). Fertilisation will result in either a normal zygote or a trisomic zygote like the parent. Both types of zygote should, in theory at any rate, appear in equal numbers. If a triso-

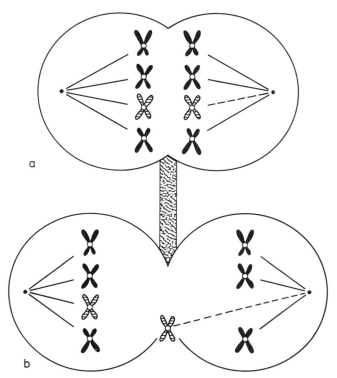

Fig. 44. Anaphase lag: faulty migration of the chromosomes at anaphase may lead to one being lost from one of the cells. If the cell is a gamete fertilisation will supply one of the pair, but the zygote will have one of the pair only and will be monosomic for the chromosome in question.

mic individual should be fertile this could happen, and has happened, though rather rarely.

At Both Meiotic Divisions

Suppose there is primary non-disjunction at the first meiotic division (Fig. 47b1 and b2); there will be two representatives of the chromosome pair in this secondary oocyte, none in the other. At the second meiotic division (remember, it is comparable to mitosis) each chromosome divides into two. Now, at the metaphase of the second meiotic division, we have four chromosomes in one cell (Fig. 47c). Depending on how those chromosomes migrate at the second anaphase, we can have gametes with four, three, two, one or no chromosome representative

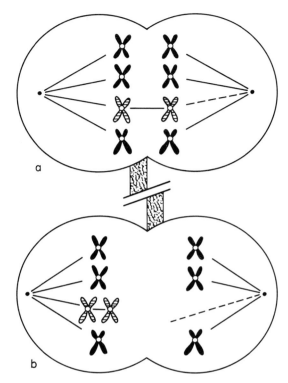

Fig. 45. Primary non-disjunction: at anaphase the two homologous chromosomes move together. One cell lacks any representative of the pair, the other has two representatives. Fertilisation of such gametes will result in monosomic and trisomic zygotes.

(Fig. 47d). Thus one could have zygotes with 'pentasomy', 'tetrasomy', trisomy, normality or monosomy of a chromosome. One can see such polysomies, though very rarely.

Mosaicism

It was stated earlier that meiotic errors are mostly responsible for chromosome disorders, and that is true, but what about the possibility of, let us say, non-disjunction at a mitotic division in the zygote? Suppose an error should be made at the first of all the many mitotic divisions, immediately after conception. Two cells with disparate, but both abnormal, chromosome complements would be formed (Fig. 48a). Descendants of those cells would be expected to perpetuate the error from generation to

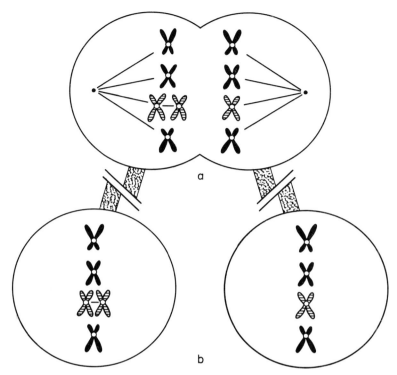

Fig. 46. Secondary non-disjunction: when a trisomic cell divides one daughter cell will have two of the three chromosomes and the other one. A trisomic parent would have (in theory, at any rate) equal numbers of trisomic and normal offspring.

generation. In fact, quite probably the monosomic stem-line would die out and the embryo, fetus and baby, if it survived to be born, would have a uniformly trisomic complement.

But suppose a non-disjunction happened just a little later, let us say, at the second of the mitotic divisions of the zygote. We would have, entering that division, two cells of normal complement, both with 46 chromosomes (Fig. 48b). Non-disjunction of one of those cells would give three types of cells with, respectively, 45, 46 and 47 chromosomes. The monosomic line would probably die out; the other two might well be perpetuated together as development of the embryo proceeds.

This embryo, fetus and baby could be a mixture, a mosaic, with a 46/47 chromosome complement.

The proportion of abnormal cells derived in this way would depend

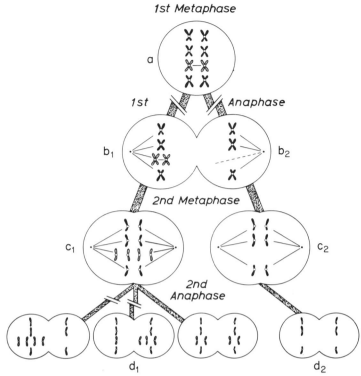

Fig. 47. Non-disjunction at both meiotic divisions; gametes could have four, three, two, one or no representative of a chromosome.

upon how many normal divisions had occurred before the error. If many normal cells had accumulated before one cell erroneously divided, the proportion of abnormal cells could be few, even trivial. It is quite possible that we are all mosaics of some degree.

The decision for the cytogeneticist as to whether a patient from whom a culture has been made is or is not truly a mosaic can be most difficult. Has the aberrant stem-line of cells been truly present in the patient—or might it have arisen by a mitotic error in the tissue culture? If the same error involving the same chromosome is present in two separate tissue culture flasks it makes a laboratory artifact unlikely. If the clinical condition of the patient is compatible with mosaicism for a chromosome abnormality that strengthens the evidence against an artifact. It is especially in prenatal diagnosis that decisions, which may involve life or death for the fetus, can be most difficult and distressing.

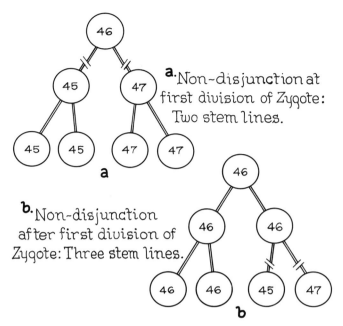

a. Non-disjunction at first division of Zygote: Two stem lines.

b. Non-disjunction after first division of Zygote: Three stem lines.

Fig. 48. Mosaic formation by non-disjunction at mitosis; two, three, or more stem-lines of cells can result, depending on the point at which the non-disjunction happens.

In mosaicism the proportion of normal to abnormal cells may vary from one tissue to another depending on how far tissue differentiation in the embryo had reached before the mitotic error took place. One tissue, blood cells, could have an entirely normal chromosome complement but skin fibroblasts could show trisomy. We will illustrate this by a case description later.

Mosaicism, then, can be of two types. All tissues can show a mixture of cells, or one tissue can show one type of cell uniformly, another tissue another.

Chimaerism

This is mentioned only to avoid confusion. Chimaerism is not mosaicism. A chimaera is an individual who has gained a stem-line of cells other than his own from another zygote. Blood cell precursors of non-identical twins may interchange through the placenta. Two cell lines can be established in one individual. Two strains of blood cells may result.

I have seen XY male cells, transfused into a female baby XX, become established and perpetuated as a chimaerism. A recipient of a donated and transplanted organ could be called a chimaera; but mosaicism must stem from a single zygote.

Causes of Chromosome Abnormalities

We are sadly ignorant of the causes of chromosome anomalies despite their frequency and importance. We know that the trisomies have a relationship to maternal age. While they do occur quite frequently in the offspring of young mothers the risk of affected children rises dramatically in women of riper years. Presumably the effect is related to the very long prophase of female meiosis and gametogenesis. There is a very slight increase with much advanced paternal age, but it scarcely reaches clinical importance. It has been said that only when father reaches age 70 is the risk comparable to that of a mother age 35. The headline 'Down's linked to old dads', scarcely seems justified!

Radiation

There seems to be no doubt that ionising radiation, X-rays in particular, can cause chromosome breaks and rearrangements in experimental animals, but mice are not men; fruit-flies even less so. It has been said that man seems peculiarly resistant to radiation genetic damage.

That is not to say we should use X-rays freely and frivolously; far from it. Evidence, to be sure, is conflicting, but some good studies have shown that such radiation to the gonads as may be acquired from pyelograms, cystograms, contrast enemas, myelograms of the lower spine and such diagnostic tests may increase significantly the risk of trisomy 21. If X-rays can indeed cause chromosome errors by mutations, one must remember that X-rays are cumulative. Time does not erase an X-ray dose. No X-ray should be ordered merely 'for the sake of completeness', not even in a child.

Despite these anxieties, there seems to be no real evidence that therapeutic radiation for, let us say, Hodgkin's disease increases the risk of offspring with chromosome anomalies. Such radiation may cause infertility, but not an increased risk of defective children.

Despite all the alarming publicity, there is no scientific evidence that exposure to computer terminals increases the risk of chromosome anomaly.

The answers, surprisingly, are not all complete. One can only say that X-rays should be regarded as very suspect and treated with respect in those of reproductive age—and younger.

Ultrasonography and NMR

There is no evidence whatever that ultrasonography can cause chromosome anomalies in a clinical situation. We do not yet know about nuclear magnetic resonance. There is no reason to suppose it will be mutagenic; but we do not know.

Pollutants

One hears a lot about environmental pollutants—Agent Orange and pollution of the Love Canal in New York State by chemical wastes—but, truth to tell, there is but scant evidence that exposure to such pollutants does cause chromosome mutations and defective children. That legal judgement has been given against a chemical company and enormous compensations awarded is not proof that harm has in fact been done.

Drugs

What about drugs? The anticancer drugs, the antimetabolites, certainly are suspect, but there is no solid evidence that a parent who has received chemotherapy for cancer is at greater than general risk of having a child with a chromosome anomaly. He or she may be infertile but that is another matter. There is no evidence that other drugs, be they medicinal or recreational, increase the risk of children with chromosome anomalies. In brief, we know very little about why chromosome disorders come about.

Nomenclature

It is convenient to have some short way of describing the chromosome complement. It is too cumbersome to write 'there is deletion of the short arm of chromosome 5 in this retarded boy' or 'there is a balanced translocation between the long arm of chromosome 13 and the short arm of chromosome 3 in this mother of this retarded girl'. There are some variations between writers, but there is much agreement.

First comes the total number of chromosomes that can be counted as individual entities. Two chromosomes stuck together count as one. There will be, then, to start a description, a number: 45, 46, 47, even, maybe, 92.

Then comes an account of the sex chromosomes: 46,XY, a normal male; 46,XX, a normal female; 45,X, a female with one X missing; 69,XXX a triploid female, three of all chromosomes.

The short arms and long arms are indicated by p and q respectively. A deletion is indicated by $-$, an addition by $+$. An isochromosome is signified by i. The loss of the long arm of one X would be written, 46,XXq$-$; an isochromosome of the long arm of one X would be 46,Xi(Xq); of the short arm 46,Xi(Xp).

Autosomes are not mentioned unless they are abnormal: 46,XY presumes 44 autosomes, but the loss of the short arm of 5 would be indicated thus: 46,XY,5p$-$, or a gain to the short arm of 3 would be 46,XY,3p$+$.

Mosaics are described by giving the different complements separated by a stroke, /. One might have 45,X/46,XX or 46,XY/47,XY,8$+$, the latter indicating mosaicism of normal cells and those with trisomy 8. Multiple stem-lines can be shown: 45,X/46,XX/46,XX/47,XXY.

Translocations are indicated by t: 46,XY,t(3p;4q) indicates a balanced translocation between the short arm of 3 and the long arm of 4. If two whole chromosomes were to join together, let us say, a translocation between chromosomes 14 and 21, one might write it, 45,XX,t(14;21).

Inversions are indicated by inv. The chromosome involved would be indicated and the precise segment involved might be indicated by the arm, the region, and the band numbers (see Fig. 29); 46,XY,inv(6)(q16q22) would be a paracentric (not involving the centromere) inversion about the middle of chromosome 6. A pericentric inversion (involving the centromere) might be written 46,XY,inv(7)(h12q11).

Deletions of part of a chromosome may be signalled by del. The site of the deletion and its extent might be shown by the arm, region and band signs: 46,XX,del(7)(q1–ter) would indicate a loss of chromosome material from about three-quarters of the way down the long arm to the end, ter or termination.

Complex chromosome rearrangements can become very difficult to reduce to a shorthand. If in doubt, talk to the cytogeneticist and ask him to explain.

From the grammar of cytogenetics we can now proceed to clinical considerations without stopping too much along the way for explanations of terminology.

CHAPTER 6

The Incidence of Chromosome Disorders

If you might think, from contemplating the exponential population explosion of mankind, that reproductive performance in man is highly successful, you would be quite wrong. Making babies is something mankind does very poorly. It is man's reproductive urges, his success in reducing infant mortality and his ingenuity and ruthlessness in sequestering the habitats and environments of other species that have led to his spectacular multiplication.

Human fecundibility, the probability of producing a liveborn infant for each opportunity to do so, is very low. One study suggests that 62 per cent of implanted embryos are lost; another study claims that as many as 78 per cent of conceptions come to nothing. It appears that intrinsic defects in the zygote and embryo are the main causes of embryo loss and spontaneous abortion. Only rarely can maternal factors be held to blame.

The Embryo and Fetus

Much of this pregnancy wastage is due to chromosome disorders. Indeed it is no exaggeration to state that, apart from senescence and its consequences, these vast genetic disasters are the commonest cause of death of mankind—of which the fetus surely is a member.

The great majority of spontaneous abortions are of fetuses of a developmental age of 8 weeks or less, though retention of the dead fetus may delay its expulsion until several weeks later. Of aborted fetuses with such early death, about 70 per cent have a recognisable chromosome abnormality. The rate of chromosome anomalies in all fetuses aborted in the first trimester is about 50 per cent. Beyond 16 weeks of pregnancy, only about 10 per cent of spontaneous abortions are of a fetus with a chromosome disorder.

With this diminishing incidence of chromosome disorders with advancing gestational age it is reasonable to extrapolate backwards to very early pregnancies, to zygotes and embryos that come to nothing so early that pregnancy and miscarriage are not recognised for what they are. It is an inescapable conclusion that very, very many conceptions end as mere 'blighted zygotes' due to chromosome disorder. Indeed, it must be so.

About 60 per cent of the aborted fetuses with chromosome anomalies have an autosomal trisomy of one kind or another. All except trisomy of chromosome 1 have been observed. Some autosomal trisomies are only very rarely observed in aborted fetuses, and yet we have no reason to suppose that these rarely observed trisomies do not happen. Autosomal monosomies are extremely rare.

Since monosomy is an essential corollary of trisomy (see Fig. 45) there must be, for every trisomy, a monosomy but they are not found in recognised miscarried fetuses. There must be blighted zygotes of equal number to the recognised trisomies. Moreover, the rarely observed but probably commonly occurring trisomies, of chromosomes 3, 9, 17 and 19 for example, very probably result in blighted zygotes—as must their counterparts, the monosomies. The chromosome disorders must indeed be very, very common and must very commonly be lethal to mankind.

Let us be thankful that they are lethal. Let us not weep over a miscarriage. Nature—and her 'quality control'—knows best what should or should not progress to viability. Liveborn babies with the chromosome disorders are those that have slipped past the quality control either by some failure of that control or because the control, however it may operate, did not recognise a potential defect as being sufficiently severe as to merit rejection.

What are the most common chromosome anomalies in the aborted fetus? Figures vary somewhat from one study to another. Figure 49 is a fair consensus. It is an odd distribution. One might well imagine that triploidy and tetraploidy would be well represented, but it is difficult to see why the sex chromosome monosomy, 45X, should so frequently be aborted when those very few that come to be liveborn have so little malformation and functional handicap. What is so bad for the fetus in being 45,X? It is quite understandable that the quality control would accept the sex chromosome trisomies for onward development. They have no malformations and few adaptive disadvantages in life.

Trisomy 16, the commonest trisomy in miscarriages, has never been seen in a liveborn baby. What can this mean? It must be that it is benign enough that the conceptus is not blighted as an early zygote but is damaging enough that, by a few weeks old, severe defects are perceived by the quality control. The same can be said of trisomy 22. Trisomy 21 puts the embryo and fetus in a most unfortunate position. It is harmful

Chromosome Abnormality in 50%

Polyploidy: Triploidy, Tetraploidy ———————— 15%

Autosomal Trisomies ————————————— 60%
　　　　　Trisomy 16 ——— 30%
　　　　　Trisomy 21 ——— 10%
　　　　　Trisomy 22 ——— 10%
　　　　　Other trisomies — 10%

Sex Chromosome Trisomies (XXX, XXY, XYY) —— 0%

Autosomal Monosomies ————————————— 0%

Sex Chromosome Monosomy (45, X) ——————— 15%

Unbalanced Translocations ———————————— 5%

Balanced Translocations ————————————— 0%

Miscellaneous ————————————————— 5%

Fig. 49. The incidence of chromosome abnormalities in spontaneously aborted recognised pregnancies.

enough that its effects lead many, indeed the majority, of its owners to be rejected as miscarriages, but it is benign enough that not all trisomy 21 fetuses are rejected. Some, a few among the many, come to be born alive, most often to a lifetime of handicap—Down's syndrome.

Because it is such a biological curiosity, one has to mention molar pregnancy, hydatidiform mole. The cells in this odd 'tumour' have only paternal chromosomes. It is believed that these moles arise from fertilisation of an 'empty egg', an egg without chromosomes: an ovum, that is, perhaps, the corollary of one that might, if fertilized, have become a triploid zygote.

Habitual Abortion

Having indicated that spontaneous abortion is very frequently due to fetal chromosome anomaly, one can deduce that, among women who have a propensity to miscarriage, there might also be a propensity to have zygotes with chromosome anomalies. Older woman indeed do

have a higher rate of spontaneous abortion than young ones because in
them non-disjunction at gametogenesis is more likely to occur. But some
woman of any age can be at especial risk of a zygote with an unbalanced
and potentially lethal chromosome complement. If one or other member
of a couple has such a balanced translocation as is shown in Fig. 42, there
will be a 50:50 chance that his gametes will have an unbalanced genetic
endowment: a 50:50 chance that the zygote will have an unbalanced
genome and perhaps, for that reason, be aborted.

There are, of course, many and varied reasons, gynaecological and
otherwise, why a woman might repeatedly miscarry, but, in the absence
of some valid explanation, most obstetricians and geneticists feel that a
series of three spontaneous abortions deserves most careful chromosome
studies of both partners, looking especially for balanced translocations as
an explanation of these pregnancy disappointments. The yield of such
explanations is not great: maybe in 7 per cent of such couples an entirely
unsuspected chromosome abnormality comes to light, most usually in
the female. But it is well worth a look.

Chorion Villus Sampling and Amniocentesis

We cannot know what an unselected sampling of fetuses by these procedures would reveal. Such tests are done for some reason, because
there is an increased risk, for one reason or another, of a chromosome
anomaly.

In women at a 0.3 per cent risk (on account of their ages of over 35
years) of having liveborn babies with Down's syndrome, amniocentesis
at 16 weeks would show a slightly greater prevalence than 0.3 per cent
of trisomy 21 among their fetuses, a prevalence of 0.4 per cent. Why is
there this difference? It is, of course, because, even beyond 16 weeks'
gestation, about 1.5 per cent of pregnancies are spontaneously aborted,
and some of these abortions will be of fetuses with trisomy 21. The
quality control was late to act.

One can, I think, expect to find a much greater disparity between the
prevalence of chromosome disorders at chorion villus sampling and the
prevalence among newborns. It would not surprise me if women with a
0.3 per cent risk of a liveborn baby with Down's syndrome were to be
found to have, in chorion villus samples, a prevalence of fetuses predes-
tined to Down's syndrome of many times the 0.3 per cent risk. One
would expect many spontaneous abortions of Down's syndrome fetuses
between the time of CVS at 9 weeks' gestation and delivery at term.

CHROMOSOME ABNORMALITIES

INCIDENCE IN NEWBORN INFANTS
(Approximate and rounded figures)

47,21+ ——— 1 : 800, both **sexes**

47, xxy ——— 1 : 750 males

47, xyy ——— 1 : 750 males

47, xxx ——— 1 : 1000 females

47,18+ ——— 1 : 3000 both sexes
(4 females/1 male)

47,13+ ——— 1 : 5000 both sexes
(1.5 females/1 male)

45, x ——— 1 : 7500 females

Balanced Translocations   –   1 : 500 both sexes

Unbalanced Translocations   –   1 : 2000

Significant inversions, deletions +

Miscellaneous ——— occasional

Fig. 50. *The incidence of chromosome abnormalities in liveborn neonates.*

Liveborn Babies

Total born

Liveborn newborns with chromosome anomalies are the few survivors who have escaped rejection among the many defective starters along the road to life. Of lifeborn babies born, about 1 in 200 has a chromosome anomaly (Fig. 50).

Down's syndrome, trisomy 21, is the most common: about 1 in 750 of all babies born, to all mothers, of all ages in all countries. Males are slightly more commonly affected than females. All other autosomal trisomies are very rare. Trisomies 18 and 13 occur with incidences of about 1 in 3000 and 1 in 5000 respectively. Trisomy 8, while usually a mosaicism, has been reported as a uniformly present complement. No other trisomies of entire autosomes have ever been observed in a liveborn neonate. Their effects must be too catastrophic. Trisomy 22, diagnosed in the days before sophisticated staining techniques led to correct identification, is not now considered to be an entity in liveborn newborns.

The sex chromosome anomalies XXY, XYY and XXX are rather common with incidences of 1 in 750, 1 in 750 and 1 in 1000 respectively. It is, perhaps, not surprising that a fetus with one of these complements escapes rejection. As we have shown (see Fig. 17 and the text), any X chromosome in excess of one is partly inactivated as the Barr body. For that reason, the supernumerary X in an XXY or XXX chromosome complement has rather little deleterious effect. The Y chromosome having, so far as is known, one gene only, coding for the HY antigen has little or no effect when present in excess. Monosomy X,45,X despite all one hears about its phenotype, is rarely seen in liveborn babies, children and adults; 90 per cent and more, as we have noted, are rejected as miscarriages. A 45,Y chromosome complement has never been seen, not even in an aborted fetus; no doubt this is because the absence of an X chromosome and all its many genes is as disastrous to the zygote as is an autosomal monosomy.

In addition to the above aneuploidies one finds structural rearrangements, inversions and translocations, balanced and unbalanced, and one finds also partial deletions and partial trisomies.

Of Older Mothers

We have noted that the prophase of meiosis at gametogenesis in the female is very long, the age of the woman at the time of ovulation plus 6 months, and we have remarked that there is ample time for things to go wrong. They can and do with a frequency related to the woman's age. Non-disjunction and the trisomies, both of sex chromosomes and the autosomes are related to maternal age. (The XYY chromosome complement and 45,X are not related to maternal age for these are due to errors in paternal gametogenesis.)

Precise figures vary from study to study, but Fig. 51 is taken from a Canadian consensus report relating to the indications for prenatal diagnosis. One sees that the incidence of Down's syndrome and the other chromosome anomalies run parallel, though it is in fact only the trisomies, both autosomal and of the sex chromosomes, that are related to maternal age. Translocations, the XYY complement and 45,X are not related to maternal age.

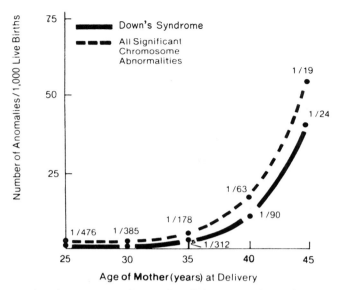

Fig. 51. The risk of Down's syndrome and of all chromosome abnormalities at different maternal ages (from the joint recommendations of the Canadian Society of Obstetricians and Gynaecologists and the Canadian Paediatric Society on Prenatal Diagnosis).

All the foregoing must surely convince the reader that the chromosome disorders are not matters that concern only the student of the recondite, but also all who are concerned with health care. They are the commonest causes of death and sickness except old age, if we accept the fetus, as we must, to be of mankind.

Recommended Further Reading

Boue, A., Gropp, A. and Boue, J. (1985). 'Cytogenetics of pregnancy wastage'. *Advances in Human Genetics, Vol. 14,* Chapter 1, Plenum Press, New York.

PART II

Clinical Considerations

CHAPTER 7

Disorders Due to the Autosomal Abnormalities

Down's Syndrome (Mongolism)

It seems appropriate, in our descriptions of the chromosome disorders to start with Down's syndrome, mongolism as it was called up to about 20 years ago when it was recognised that this name gave offence to some. Down's syndrome is the paradigm of the chromosome disorders. It is the most common, the best studied and shows best the classical features and variations of a chromosome abnormality.

In 1866 John Langdon Down, while Medical Superintendent of the Earlswood Asylum for Idiots in Surrey, England, wrote his now famous paper 'Observations on an Ethnic Classification of Idiots' in the London Hospital Reports. It opens thus:

> Those who have given any attention to congenital mental lesions must have been frequently puzzled how to arrange, in any satisfactory way, the different classes of this defect which may come under their observation. Nor will the difficulty be lessened by what has been written on the subject.

This is largely true today although more than a hundred years of intense study have passed. But it is, in fact, a result of Down's own attempt at a new classification that we have come to recognise one well-defined type of mental retardation.

> I have been able to find among the large numbers of idiots and imbeciles that have come under my observation . . . that a considerable proportion can be fairly referred to one of the great divisions of the human family other than the class from which they have sprung . . . . Several well-marked examples of the Ethiopian variety have come under my notice, presenting the characteristic malar bones, the prominent eyes, the puffy lips, and the retreating chin . . . . Some arrange themselves around the Malay variety, and present in their soft, black, curly hair, their prominent upper jaws and capacious mouths, types of the family which people the South Sea Islands. Nor have there been wanting the analogues of the people, who with shortened foreheads, prominent cheeks, deep set eyes, and slightly apish nose, originally inhabited the American Continent.

The great Mongolian family has numerous representatives, and it is to this division, I wish, in this paper to call special attention. A very large number of idiots are typical Mongols. So marked is this, that when placed side to side, it is difficult to believe that the specimens compared are not children of the same parents (Fig. 52).

He then goes on to describe the typical 'Mongolian' idiot, mentioning the broad flat face, the obliquely placed eyes and the heavy epicanthic folds. He draws attention to the thick fissured lips, the thick rough

Fig. 52. Down's syndrome; a group of unrelated patients.

tongue and the small nose. He describes the skin (perhaps he has mistakenly diagnosed some hypothyroid cretins as 'Mongolian' idiots) as having a dirty yellowish tinge and as being deficient in elasticity. He remarks, rightly, on their lively sense of humour and of the ridiculous and their aptitude for mimicry. He states that while most learn to speak, the speech is thick and indistinct. He hints at their special susceptibility to winter infections and strongly emphasises the gains that will result from efforts at training. He gives, indeed, a masterful first description of this syndrome that we have come to know so well by his name.

However, when he ventures to speculate on aetiology he goes astray. He makes the good point that 'they are always congenital idiots and never result from accident after uterine life' but he claims that 'they are, for the most part, instances of degeneracy arising from tuberculosis in the parents'. He appears to regard the condition as a reversion to a primitive mongolian ethnic stock. His son, Reginald, himself also a specialist in mental retardation, later repudiated his father's views and wrote 'It would appear that the characteristics which at first sight strikingly suggest mongolian features and build are accidental and superficial, being associated, as they are, with other features that are in no way characteristic of that race. . . .'

John Langdon Down considered that 'The Mongolian type of idiocy occurs in more than 10 per cent of cases that are presented to me'. Over a hundred years later a report from New York State gave the incidence of Down's syndrome among 12 000 retarded children as 9.8 per cent.

Following Down's description little was added along the way in the next 90 years in spite of an enormous amount of speculation. However in 1876 Fraser and Mitchell pointed out that such patients tended to be born at the end of large sibships, leading to the notion that in some way 'uterine exhaustion' was responsible. In 1909 Shuttleworth pointed out that it could be maternal age, rather than parity, that was the relevant factor.

In 1933 Penrose showed that there was indeed a relationship to maternal age, independent of parity. In 1951 he showed that a curve of the incidence of mongolism (as it was still called in those days) plotted against maternal age showed two peaks: the one corresponding to the maternal age at which the birth rate was highest (around 25 years, at that time), the other corresponding with a maternal age of around 40 years. Moreover he showed a significant lowering of the mean maternal age at the birth of the mongol child in those families where there was a mongol relative or where the mother had herself previously had a mongol child. There was a hint that there might be two varieties of mongols: those in which maternal age was relevant to aetiology, and those that were independent of maternal age (Fig. 53).

It had also become known that monozygotic, 'identical', twins were almost invariably concordant for the disease. If one was a mongol, the other was also. In dizygotic twins the concordance was not significantly greater than between non-twin sibs. The disorder appeared to be of genetic origin and not due to intrauterine environmental events. It had also been noted that in the rare event that a mongol woman had a child, the chances of a baby with mongolism or normality were approximately even.

As long ago as 1932 de Waardenburg proposed that mongolism

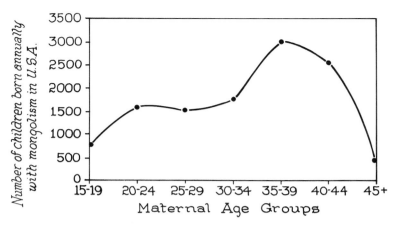

Fig. 53. The number of children with Down's syndrome born annually in the United States to mothers in the different age groups (from C. E. Benda, The Child with Mongolism, Grune and Stratton Inc., New York).

might be due to a chromosome abnormality. This was a most prescient but, at that time, unsupported suggestion.

In 1959 Lejeune, Gautier and Turpin in Paris, applying the knowledge of the normal array of human chromosomes gained 3 years before by Tjio and Levan, showed beyond doubt that the cells of mongols had an extra small acrocentric chromosome in group G, by convention number 21. They found trisomy 21 in mongols. They had three of a kind, not a pair; they had 47 chromosomes, not 46.

Soon after this Polani discovered a mongol child with 46 chromosomes—or so it seemed. In fact there was, indeed, an extra 21 chromosomes but it was joined on to, translocated onto, one of the other chromosomes. The extra chromosome was there, exerting its evil influence, just as as if it had been 'free'.

In 1961 Clarke and her colleagues discovered a child who, while having some features of mongolism, was not typical of that disorder. They found in this child two cell populations: some cells were normal, some had the extra chromosome. She was a mixture of cells: a mosaic, in this case a 'mongol mosaic'.

In 1973 Dent found another example of a child with some features of mongolism, but not all. In that child only part of chromosome 21 was present in excess.

In 1967 Lejeune and his colleagues described what might be regarded as the opposite, the '*contre* type' of mongolism: 'antimongolism'. This child had part of chromosome 21 missing, deleted from the complement. Later Al-Aish reported the very rare but not quite unique event, as later cases showed, of a baby with truly 45 chromosomes only: a monosomy of chromosome 21.

Undoubtedly many fetuses with trisomy 21 are recognised as such by the 'quality control' of the uterus. If 15 per cent of all recognised pregnancies are miscarried spontaneously and if 50 per cent of those have a chromosome anomaly, and if, furthermore, 10 per cent of that 50 per cent have trisomy of chromosome 21, one can see that many more babies with potential Down's syndrome are conceived than ever come to be born alive. Let us not weep over a miscarriage!

Incidence of Down's Syndrome

Let us now revert to the modern name for the clinical disorder characterised by trisomy of chromosome 21: Down's syndrome.

One might imagine from all the interest that has been focused on Down's syndrome in the last 100 years that it is a very common condition. But it is not. About one in every 750 babies has Down's syndrome—if you like, a single joker in 15 decks of playing cards. Congenital heart disease, all types together, is four times more common; cystic fibrosis half as common—at birth, that is. The prevalence of Down's syndrome in a school-age population is a good deal less, for about 40 per cent of Down's syndrome babies have congenital heart disease of which many die, even with modern cardiac surgery.

The condition is recognised in blacks and is well known among Asiatics, even among the inhabitants of Mongolia. The incidence is much the same the world over. Some studies, but not others, have seemed to show a variation in incidence with the season of the year. Some studies have purported to show a relationship with epidemics of viral infections in the community. One study purports to show that where sexual intercourse is infrequent, Down's syndrome is more common. At least two studies have shown a higher incidence in Roman Catholics than among Protestants, the effect being due, not apparently to childbearing continuing to a later age in the former than the latter, but possibly to the employment of the 'rhythm method' of birth control. Even concerning any association between radiation of the mother prior to pregnancy and the incidence of Down's syndrome there is disagreement. Some workers claim such association; others deny it. There certainly does not seem to have been a remarkably high incidence of Down's syndrome in the offspring of the survivors of the atomic holocausts of Nagasaki and Hiroshima. Only maternal age is irrefutably related to Down's syndrome.

As a woman gets older her chance of having a liveborn baby with Down's syndrome rises, irrespective of how many children she has previously had. A mother of, say, 23 years runs a risk of about 0.04 per cent or odds of 2500 to 1 against the misfortune. By age 30, the risk is about the same as the overall risk of Down's syndrome in all babies of all mothers of all ages, say 1:800. At age 35, the odds have shortened to 1: 300–350, about double the general risk; or, to put it in perspective, the chance of drawing a single joker from six or seven decks of cards. At age 40, the risk is about 1 per cent, or 100 to 1 odds against the catastrophe. At age 45, the risk may be as high as 4 per cent: thirty times the general risk (see Fig. 51).

Paternal age has little effect—the headline 'Down's linked to old dads!' has little substance. The father must be about 70 years of age before the risk on his account becomes substantial.

The couple who have had a Down's syndrome child, even one by the chance mishap of non-disjunction at gametogenesis, is at a greater than general risk of a repetition of trisomy 21. There seems to be in some people a predisposition to non-disjunction. We do not know why.

Until recently it has been held that it is always maternal gametogenesis that is at fault: that the non-disjunction is an error in the ovary. Such a belief is no longer tenable in all cases. It may be so in the elderly mother, in the mother of let us say 35 years of age or more, but it is not true in young couples.

Chromosomes of all people are not identical, even the chromosomes of the same homologous pair. There may be 'marker' or distinguishable chromosomes. In Fig. 54 we see that father has one 21 chromosome with distinct satellites on the short arm: mother's chromosomes do not have such satellites. The Down's syndrome baby has three 21 chromosomes: two with satellites, one without. In this case the father clearly contributed two of the three chromosomes. The error was in sperm gametogenesis. In young couples with a Down's syndrome baby such is the case quite often: say 30 per cent. There are not often marker chromosomes that can distinguish the maternal or paternal contribution—even if it were deemed wise to try and assign responsibility—but at least one can say to a couple, with truth, that either could have given the extra chromosome.

Clinical Features

It is not our purpose here to describe in detail the clinical features of the disorder. Such descriptions can be found in standard texts.

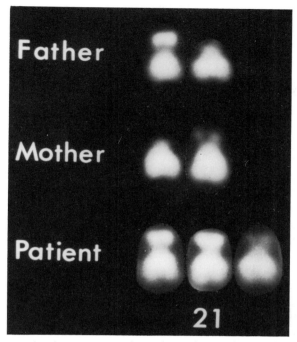

Fig. 54. A marker chromosome 21 indicates that in this case the extra chromosome 21 was contributed by the father.

The most constant and distressing feature is the universal mental retardation. The IQ varies between 30 and 65 with a mean value around 50. Whether the variation is determined by differing severity of the disease itself or whether it is much modified by environment and upbringing it is difficult to say. It seems to this author that a Down's syndrome child born to parents of high intellect, and therefore, with a high hereditary intellectual potential, is less retarded than one born to dull and backward parents. It has struck me that Down's syndrome seems to cut the potential IQ in half. No doubt environment, and perhaps special programmes of stimulation, play their part in determining the achieved IQ but even the brightest Down's syndrome child will not become a fully self-supporting citizen.

It is usually very difficult at first for parents to accept the diagnosis for, to them, the baby shows no very striking features and no evidence of retardation. These babies are at their best in infancy, but as time goes on psychomotor development falls further and further behind until by

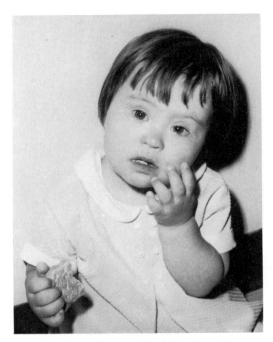

Fig. 55. A typical young girl with Down's syndrome.

the end of the first, and certainly by the end of the second year, the retardation can scarcely be denied by even the most optimistic parent.

There is one bright side to this sad picture. These are happy, affable, humorous, affectionate and often winsome children. They are not continually crying, whining, destructive and evil-tempered as are some other retarded children. They are easy to handle in the home, lovable and often much loved. If one must have a retarded child, it is best to have one with Down's syndrome.

Walking is likely to be delayed to 18 months, 2 years or even longer; but they all walk. Speech development is always slow but almost all will talk. Usually the voice is thick and indistinct; phrases and words are simple. Toilet training, like walking, takes time but almost all achieve good control. Toilet training does not seem to be a major problem.

The appearance is characteristic (Fig. 55), and yet it is as difficult to state in factual terms just what it is that makes a child recognisable as a case of Down's syndrome. Analysis of individual features does not help too much. It is the overall impression that makes for the diagnosis, in the same way as it is the overall impression that makes for the diagnosis of a pretty girl.

Rather surprisingly, there has been no certain portrayal of Down's syndrome in art down through the ages, and this has led to the question of whether Down's syndrome is of only recent appearance. But this is quite unlikely. To this author the idiot in the 16th century painting by Hieronymus Bosch of a quack surgeon removing the 'stone of folly' looks like an adult mongol.

Down's syndrome children are small. The mean birth weight is about 250 g below the mean birth weight for normal babies. Throughout life they remain short. Most are below the 25th percentile for height, and for their height they are usually overweight.

Hypotonia is striking in the babies; they are extremely floppy. Their joints may be hyperextended and their limbs placed in grotesque positions. This hypotonia persists to some degree into adult life. The adult patient has a characteristic flabby handshake. The feet are short and broad with a poorly developed arch. There is often a wide gap between the first and second toes. The head is small and rather square, foreshortened from back to front, brachycephalic. The neck is short and thick. Recently it has been recognised that some children with Down's syndrome have an instability of the upper cervical spine at the occipito-atlantoid articulation and that this can be a hazard in tumbling, wrestling and play on a trampoline. It is recommended that such activities be discouraged unless X-rays have excluded such an instability.

The eyes are set closer together than normal because of the underdevelopment of the skull. In childhood, but less so in later life, the eyes are slanted upward at the outer side. The iris is more commonly blue or grey than in the general population, and in the blue or grey-eyed patient small white triangular speckles, Brushfield spots, are often seen around the periphery of the iris. They are not diagnostic of Down's syndrome, for they can also be seen in normal people. Squints are common (Fig. 56) and in later years chronic blepharitis and lens opacities are often seen.

Much is made, too much, of epicanthic folds of skin: crescentic folds running from the upper lid to the nasal bridge. While these epicanthic folds are present in most young children with Down's syndrome they are so common in normal infants as to have little significance in diagnosis (Fig. 57).

The nose is small, and the patient is plagued with persistent obstructions and infections. The Down's syndrome child snorts and snuffles his way from one infection to the next. This is in part because of the small nasal airway but also because these children, as part of the disorder itself,

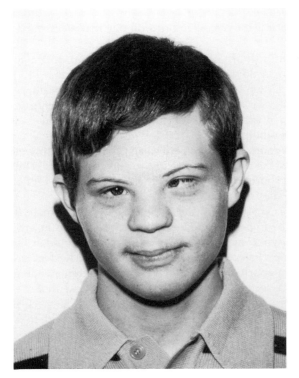

Fig. 56. An adolescent with Down's syndrome and a squint.

have some impairment of their immune system. The tongue, too large for the small mouth (which is often open because of nasal obstruction), often protrudes and becomes thickened, fissured and wrinkled: a 'scrotal' tongue. The lips are thick and fissured and in later life the lower lip often hangs down. There is often malocclusion of the teeth which, however, seem unusually resistant to dental caries. The chin is often receding and filled-in below the jaw (Fig. 58).

The trunk is not especially unusual though the nipples are rather flat. In females pubertal breast development is usually delayed and in later life the breasts are large, fat and pendulous.

Congenital heart disease is very common. About 40 per cent of Down's syndrome children have congenital heart disease. Most usually the defect is a septal defect; of those with congenital heart disease, about 40 per cent have an A-V commune defect, a cushion defect with, usually,

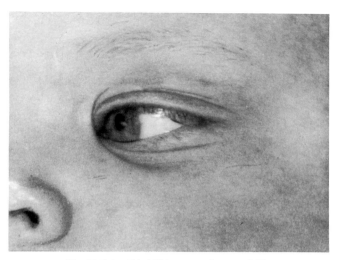

Fig. 57. Epicanthic fold in a normal young child.

a cleft mitral valve and mitral regurgitation. If there is a left-to-right shunt from any cause, the increased pulmonary blood flow seems especially damaging to the infant's lungs, probably because hypoplasia of the lungs and pulmonary blood vessels is part of the disorder itself. Increasing pulmonary vascular resistance, the Eisenmenger syndrome, and reversal of the intracardiac shunt and cor pulmonale may develop before 2 years of age. A careful watch must be kept for early evidence of increasing pulmonary vascular resistance and of increasing right ventricular strain. Such developments may call for early surgical closure of the defect, before irreversible changes make it too late.

Anomalies of the gastrointestinal tract are quite uncommon, except one: duodenal atresia. If a baby with Down's syndrome shows signs of very early intestinal obstruction it is likely that there is duodenal atresia. If a newborn baby has duodenal atresia one must think that the baby might have Down's syndrome. The male genitalia are often abnormal. The penis is usually quite small and undescended testes are very common. Male puberty is often much delayed and testicular degeneration and infertility seem to be the rule. The female external genitalia are not very remarkable though the labia majora may be prominent and cushion-like. Female puberty and menarche are usually much delayed. Menstruation is scanty and irregular; the menopause comes early. These women

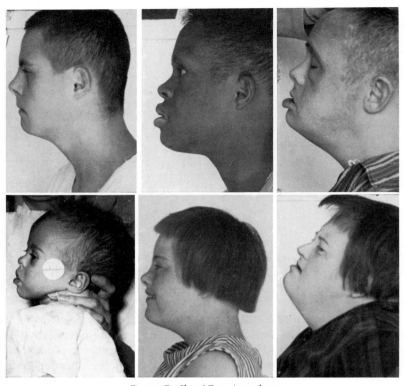

Fig. 58. Profiles of Down's syndrome.

have very low fertility but some births to Down's syndrome patients are on record. As one might expect, the mechanism of secondary non-disjunction (see Chapter 5) gives offspring with about an even chance for normality or Down's syndrome.

It is well established that the acute leukaemias are much more common, some say 15 to 20 times more common, in patients with Down's syndrome than in normals. We do not, for sure, know why. It may be that impairment of immunological surveillance allows malignant changes to go unrecognised and unchecked. It may be that the chromosome disorder itself may allow the activation of normally present pro-oncogenes into active cancer-producing oncogenes. Certainly such activation seems to result when there is reciprocal translocation involving chromosomes 9 and 22 with the formation of the 'Philadelphia chromosome' which is so strongly associated with chronic myeloid leukaemia.

We must admit that we do not know just what the underlying micro-biological defects are that result in Down's syndrome when there is triple representation of chromosome 21, nor do we know why some unfortunate persons (not only women, as we have noted) suffer non-disjunction at meiosis while others do not. There is, perhaps, a slight familial predisposition to non-disjunction. Non-disjunction trisomy 21 Down's syndrome does seem to occur more frequently in some families than chance alone would allow; we do not know why.

One thing has been noted by several workers. There is an increased incidence of thyroid dysfunctions of all kinds in parents, sibs and near relatives of patients with Down's syndrome due to non-disjunction. Even if there is no overt evidence of thyroid disease, thyroglobulin auto-antibodies are often found in close relatives. The association is not specific for Down's syndrome. An increased incidence of thyroid autoantibodies is found in close relatives of patients with other chromosome anomalies. Down's syndrome patients themselves have a higher than expected rate of the 'autoimmune diseases', insulin-dependent diabetes mellitus, Addison's disease and, especially, Hashimoto's thyroiditis. This is curious, but lacking valid explanation even though these facts have been known for nearly 30 years.

Obviously many genes, coding for many proteins and enzymes must be located on chromosome 21. One would expect increased activity if there were triple, rather than double, representation of a coding gene. Thus far only the enzymes superoxide dismutase, phosphoribasil-glycinamide synthetase and hepatic phosphofructokinase are known to be coded for on chromosome 21. In fact extra activity of these enzymes can be demonstrated in Down's syndrome.

The average expectation of lifespan is reduced in Down's syndrome. Apart from early death in many cases from congenital heart disease, there is also premature ageing especially in the nervous system. There is early decline in the already compromised intellectual function and at autopsy there are changes very similar to those found in Alzheimer presenile dementia. It seems as though in some way the chromosome abnormality hastens the genetically programmed course of 'shutting down' of the enzyme systems that is part of the normal ageing process. One notes at this point the curious fact that Down's syndrome is more likely to occur in those families in which Alzheimer's disease is clearly genetically determined. There seems to be some relationship. Be that as it may, only 8 per cent of Down's syndrome patients survive beyond 40 years of age, and only about 3 per cent beyond age 50. However, there is on record a lady with Down's syndrome who lived to be 84.

Fig. 59. Stephen C: partial Down's syndrome due to reduplication of part, but only part, of chromosome 21.

Questionable Down's Syndrome, Partial Trisomy, Mosaicism

Is Down's syndrome an 'all or none' condition? Are there degrees of Down's syndrome? In general the condition is absolute with, to be sure, some variation in intellect depending on hereditary endowment and environmental influences. But there are exceptions. There are partial trisomies of chromosome 21 and partial expression of the features of the syndrome. It seems as though the segment of chromosome 21 that in triple dose results in the classical features of the syndrome is the end of the long arm of the chromosome: 21q22 — ter (see Chapter 2 and Fig. 29).

Stephen C had for a long time been regarded by experienced staff members of a hospital school for the retarded as a case of Down's syndrome. His appearance was strongly suggestive but not quite typical (Fig. 59). Chromosome analysis of several tissues showed the normal number of chromosomes, 46. However very careful study of the chromosomes has shown that there is reduplication of the tip, and only the tip, of the long arm of one of his 21 chromosomes: a 'terminal reduplication'. He is a 'partial Down's syndrome' because he has only a trisomy of part of chromosome 21.

Kevin Mc (Fig. 60) was not recognised as unusual in early infancy but, when at 2 years of age he showed some developmental delay, he came

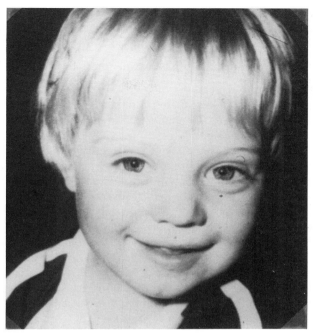

Fig. 60. Kevin Mc: Down's syndrome mosaic.

under closer scrutiny. His chromosomes were reported by the laboratory as normal in lymphocytes cultured from a blood sample. There the matter rested for 2 or 3 years until continuing developmental delay led to reassessment. Further study showed a small proportion of cells trisomic for chromosome 21 in the lymphocytes, but cultured skin fibroblasts showed 67 per cent to be trisomic. He is a case of partial, but recognisable, Down's syndrome; partial because his cells are part trisomic, part normal. He is a Down's syndrome mosaic. One estimate is that about 2 per cent of recognisable cases of Down's syndrome are mosaics.

This author sees many people in the street, in cinema queues and all over the place, apparently functioning quite normally in society and yet with a decided 'Downsy' look to them. How one would love to look carefully at their chromosome complements, if possible in several tissues! Are they Down's syndrome mosaics? Who knows. But if they are, and if some cells in their gonads were to be trisomic, they might be at exceptional risk of having Down's syndrome children by the mechanism of secondary non-disjunction (see Fig. 46). One can never know if someone might be a chromosome mosaic of some degree until every last body cell has been sampled.

Antimongolism, Monosomy 21

Having said that the expression of Down's syndrome can be dependent on the dose of extra genetic material, either by partial trisomy or by mosaicism, it is legitimate to enquire what might be the result of a lessened amount of chromosome 21.

As we have noted the formation of a ring chromosome is usually, perhaps always, associated with some loss of genetic material. De Grouchy and Turleau in their recent edition of the *Clinical Atlas of Human Chromosomes* cite eleven cases of r(21) in which several features seem to be the contratype of Down's syndrome. Microcephaly is constant but the occiput is protuberant rather than flat, as in trisomy 21. The nose likewise tends to protrude rather than be small. The eyes may have an 'antimongoloid' slant. The ears are large rather than small and the external meati wide rather than narrow, as they are in trisomy 21. Micrognathia seems to be constant. Malformations of many kinds are usual. Interestingly pyloric stenosis is unexpectedly common whereas it is said to be quite uncommon in babies with Down's syndrome.

Here the dissimilarities end. These patients with deficiency of chromosome 21 show, if they live long enough—and many do not—very severe mental retardation.

Causes of complete monosomy of chromosome 21 are poorly authenticated and of somewhat dubious validity. Very probably monosomy 21 is incompatible with fetal development, as all autosomal monosomies are. De Grouchy cites only a single example of mosaicism of monosomy 21 and states that the phenotype was very similar to the r(21) syndrome described above.

All one can say is that there do seem to be, but very rarely, partial deficiencies of chromosome 21, with clinical features that could legitimately be described as 'antimongolism'.

Dermatoglyphics and a Dermatoglyphic Nomogram

In relation to Down's syndrome we may usefully consider here a form of clinical examination that can be of value in diagnosis, especially if the result of a chromosome study may be long delayed. This is a study of the fine dermal ridge patterns, the 'fingerprint' patterns on the finger tips, the palms and the soles of the feet. It was found by Cummins in 1939 that the patterns of these fine ridges, the dermatoglyphic patterns, differ in cases of Down's syndrome from those found normally.

If you look at your finger tip with a strong lens (an electric auriscope with the snout removed, or an ophthalmoscope with a + 15 or + 20 lens will do quite well) you will see the fingerprint ridges. You will see that in most patterns there will be one or more points where the ridges meet together forming a triangle or 'triradius' (Fig. 61).

Fig. 61. Triradius in a dermatoglyphic pattern.

The patterns themselves are of three kinds: there may be, not very commonly, simply an arch-shaped arrangement of ridges without any triradius; or there may be a pattern with one triradius making a 'loop', which may be either big or small or may be 'open' to the radial or ulnar side of the hand. A radial loop is open to the radial side, with the triradius on the ulnar side; an ulnar loop has the triradius on the radial side and is open to the ulnar side (Fig. 62). The third pattern is the 'whorl' which may be quite simply a pattern of concentric circles or may be complex. A whorl, whatever else there may be in the ridge configurations, has two triradii (Fig. 63).

It has been found that in normal people some patterns are more common than others. If one takes the patterns on the ten digits one finds a frequency thus: loops, both ulnar and radial, 70 per cent, whorls 25 per cent, arches 5 per cent. The frequencies do not hold good for the digits individually. Each digit has its own individual frequency. For example, while on every digit loops are normally abundant, their frequencies range from 85 per cent on the little finger, digit V, to 34 per cent on digit II. There is, moreover, a difference between the two hands. Whorls are

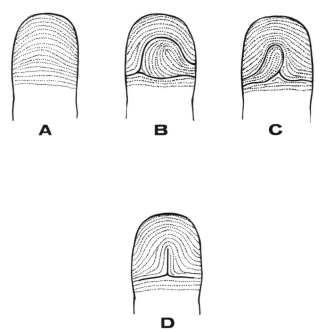

Fig. 62. *Dermatoglyphic finger patterns: A arch; B large radial loop, if in the right hand; C small ulnar loop on the right hand; D tented arch.*

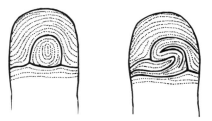

Fig. 63. *Whorl patterns, simple and complex.*

more likely to be found on the thumb, digit I, of the right hand (39 per cent) than on the thumb of the left (31 per cent). From a study of large populations tables have been constructed giving the frequencies of the patterns on the separate digits of both hands in normal person.

The relevance of this to us is that the frequency distributions in Down's syndrome are different. For example, let us take the index finger, digit II, of the left hand. We see quite a different frequency distribution

PATTERN	DIGIT II, LEFT HAND	
	Down's	Normal
Whorl	1 1.9	33.4
Ulnar Loop	82.4	36.3
Radial Loop	2.3	1 9.4
Arch	3.4	1 0.9

Fig. 64. Frequency of the four digital patterns on the index finger of the left hand of normals and Down's syndrome patients.

between normal persons and Down's syndrome patients: whorls are three times more common in normals, radial loops eight and a half times more common, arches three times more common in normals than in Down's syndrome, but ulnar loops are less common in normals than in Down's syndrome: two and a quarter times less common. We might say, then, that a radial loop on the index finger of the left hand gives odds in favour of normality of 8.5 : 1; or an ulnar loop odds of 2.27 : 1 in favour of Down's syndrome (Fig. 64).

If you now look at the palm of your hand you will see a number of points where ridges meet forming triradii. There will be one at the base of each finger, a, b, c, d (Fig. 65). There will also be one, the axis triradius, somewhere in the palm, usually normally back towards the wrist but sometimes, unusually in normals, shifted distally. If one makes a line— and this can be drawn on the palm with a ballpoint pen—joining the a and d triradii to the axis triradius one makes an angle, the 'atd angle', which will usually be quite acute, but may be more or less obtuse, depending on the width of the hand and the position of the axis triradius (Fig. 65). Wide atd angles are more likely to be found in Down's syndrome (and in some other chromosome anomalies) than in normals; the more obtuse the angle, the greater the atd angle, the more likely is it that the patient will have Down's syndrome.

If we now look at the hallucal area of the foot, the 'ball' of the foot, we

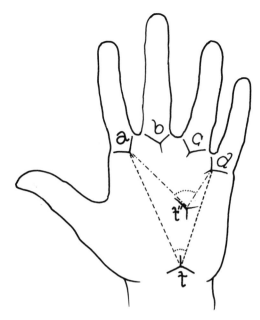

Fig. 65. The atd angle: normal t, abnormal t″.

see somewhat similar patterns to those on the fingers. The most common is a loop, a pattern with a single triradius. Next most common are concentric circles with two triradii, the whorls. The loop will be a 'distal loop', open distally between the great toe and second toe with the triradius on the medial side of the foot. The loop may be large, most commonly, or small (Fig. 66a and b). Together loops make up about 50 per cent of hallucal patterns, with large loops making 40 per cent, small loops 10 per cent. The larger the loop the more likely is normality. The smaller the loop the more likely Down's syndrome. If the loop is so very small that there is no real loop at all, only an arch open to the tibial side (a 'tibial arch' or TbA; Fig. 66c) a very high loading in favour of Down's syndrome can be given, for it is about 150 times more likely to be found in Down's syndrome than in normals. A tibial arch on both feet makes normality quite unlikely.

There is a rather complicated way of using data from ten digits, both palms and both feet, to derive a 'Ford Walker Index' related to the probability of Down's syndrome. A much simpler and valid method uses only four observations: the patterns on both index fingers, the pattern on the right hallucal area and the atd angle on the right palm. UL is an

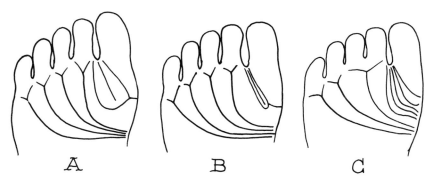

Fig. 66. Hallucal area of foot; A shows large distal loop; B small distal loop; and C a tibial arch.

ulnar loop, RL a radial loop; W is a whorl, A an arch. TA is a 'tented arch' (see Fig. 62). LDL and SDL are large and small distal loops respectively. FL is a loop open to the fibula side, a rare finding. TbA is the sinister tibial arch.

Using the nomogram (Fig. 67) one selects the right hallucal pattern and joins it to the atd angle (1). One next selects, and joins, the index finger patterns (2). Then one joins the points where lines (1) and (2) cut lines A and B (3). Finally one sees where line (3) cuts the 'diagnostic index line'. In the example shown we have a tibial arch, a rather large atd angle, an arch on the right and an ulnar loop on the left index finger— and we have a very strong indication that such a patient has Down's syndrome.

Surprisingly it works! There are few doubtful cases, but there is one real problem. In a newborn baby, suspected of Down's syndrome, about whom one might wish to voice these suspicions to the parents before the chromosome analysis is complete, it can be very difficult accurately to read the dermatoglyphics. In a baby of a few weeks old it is quite easy, but the urgency of diagnosis will be less.

Creases

Although the study of the tiny dermal ridge patterns is of quite recent date (Sherlock Holmes made no use of them in his detective endeavours) the transverse palmar crease, the 'simian' crease (Fig. 68) has been of interest to clinicians for some time. It has some diagnostic value in Down's syndrome for it is present on one hand or the other—and often both—in about 40 per cent of Down's syndrome patients compared

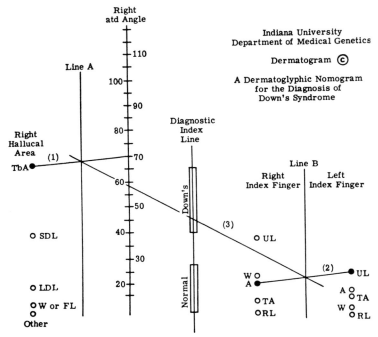

Fig. 67. Dermatoglyphic nomogram for the diagnosis of Down's syndrome. Dermatogram with permission from the Department of Medical Genetics, Indiana University

with 4 per cent in normal people. Perhaps too much is made of the transverse palmar crease. It certainly is not diagnostic of Down's syndrome, or indeed of a chromosome anomaly. Apart from the fact that it is found in normal people it is found quite commonly in patients with congenital heart disease and other malformations, including the defects due to prenatal rubella infection of the fetus.

A single crease on any digit, except of course the thumb, is much more suggestive of a chromosome anomaly. A single-digit crease, usually on the little finger, is very uncommon indeed in normal persons. It is seen in about 20 per cent of cases of Down's syndrome. A single-digit crease deserves a chromosome analysis. I do not know the significance of three digital creases. I have seen this only once—in one of my paediatric colleagues. He seems normal enough!

Cytogenetic Mechanisms in Down's Syndrome

Cytogenetically speaking there are several varieties of Down's syn-

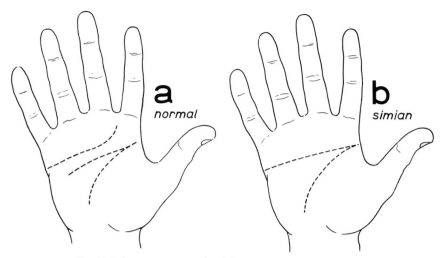

Fig. 68. Palmar creases, normal on left, transverse (simian) crease right.

drome. There is the 'regular' trisomy 21 due to non-disjunction in one or other of the parents. There is the unbalanced translocation trisomy in which one or other of the parents carries a balanced translocation involving chromosome 21 and, commonly, chromosome 14. Then there is the unbalanced translocation trisomy with an extra chromosome 21 attached to another chromosome, again usually 14, but with neither parents being balanced translocation carriers. These are the *de novo* (in that individual uniquely) translocation cases of Down's syndrome. Then, again, there are those who have Down's syndrome because of reduplication of the whole of the long arm of chromosome 21 by the mechanism of isochromosome formation (see Fig. 43); and then there are the Down's syndromes due to mitotic non-disjunction of chromosome 21 in the zygote, after conception (see Fig. 48).

The frequency with which the various types are found depends on the method of selection of the cases studied. If the Down's syndrome children of elderly mothers were to predominate in the sample almost all would be found to be of the regular trisomic variety due to non-disjunction, related to maternal age. The older the mother the more likely is she to have a baby with Down's syndrome, and the more likely it is that her Down's syndrome baby has its trisomy by non-disjunction. If on the other hand one selects for study the Down's syndrome babies of young mothers a significant number will be found not to be of the regular trisomic variety though, even so, the number of the regular variety will greatly outnumber the cases due to mechanisms other than non-disjunction. Again if Down's syndrome babies are selected for study from those families where a sib or close relative has Down's syndrome a very significant number will be found to have the Down's syndrome as an unbalanced translocation trisomy of chromosome 21.

In a large series of 453 studied by Uchida, 13 had a translocation trisomy, but only four were due to a translocation-carrying parent; in nine the translocation trisomy had arisen *de novo*. There were, rather surprisingly to this author, five mosaics, one by isochromosome formation and one with a double trisomy: 48,XXY,21 +—Down's syndrome plus Klinefelter's syndrome. It is fair to say that in all Down's syndrome babies, of all mothers of all ages, 96 per cent are of the 'regular' trisomic type.

The 'Regular', Trisomy 21, Down's Syndrome

As we have seen in an earlier chapter, the gametes, sperm and ova, are formed by a reduction division at meiosis. It will be recalled that this takes place in two stages, the first and second meiotic divisions. At either of these divisions, and usually at the first, there can be incorrect segregation and migration into the daughter cells of that division. Non-disjunction may occur as illustrated in Fig. 45. One gamete will have a chromosome too many, both of the pair; the other will have one too few, with no representative of the non-disjuncting pair.

Fertilisation of these gametes must result in abnormal zygotes. One will have three representatives of the relevant pair, the other but one. There will be trisomic (Fig. 69) and monosomic zygotes. If chromosome 21 is the pair involved, the trisomic zygote is destined to be a Down's syndrome baby—unless it is rejected as a miscarriage along the way. The monosomic zygote will, with perhaps very, very rare exceptions, die in very early embryonic life and be reabsorbed—not even recognised as a pregnancy.

We just do not know what causes these non-disjunctions. We do not really know why there is a relationship to maternal age. It is too facile to speak of this as a chance mishap. Mishaps must have a cause. We can, to be sure, postulate that there is a 'shelf-life: to the developing ovum: that it might be labelled, like dairy products in a grocery store, "best before . . ." '. We can further postulate that it is the long, long prophase of the first meiotic division, the state of suspended activity before ovulation completes that division that leads to non-disjunction; it is as though impatience to make its ovulatory debut leads to the ovum undergoing

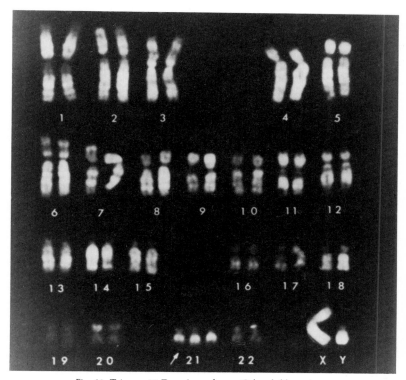

Fig. 69. Trisomy 21 Down's syndrome; Q-banded karyotype.

non-disjunction. This may be the truth but there are other possibilities. Maybe with advancing years (and sometimes in young mothers for other reasons) there is delay in fertilisation because passage of the completed ovum into the fallopian tube is sluggish. Maybe the ageing uterus fails in its 'quality control' and does not reject so readily an embryo with an abnormal chromosome complement. We just do not know.

And what about paternal age? Within practical limits there is no increased risk of a Down's syndrome baby to fathers of riper years. It has been estimated that father must be 70 years of age and more before the risk is as great as that given by a maternal age of 35.

De Novo Translocation Down's Syndrome

While relatively rare, about 2 per cent of all Down's patients, the *de novo* translocation is the next most common.

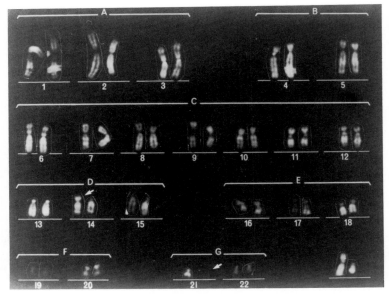

Fig. 70. Karyotype of translocation trisomy Down's syndrome.

If one counts the apparent number of chromosomes there will seem to be 46, the normal number. This is not so; in this condition there are really 47 but one, chromosome 21, is attached to, translocated to, another; usually it is to chromosome 14, but it may be to its homologue, the other chromosome 21. In the *de novo* translocation there is no abnormality in the parents. This is an event unique to that individual. The chromosome array, the karyotype, is identical with that of the inherited translocation (Fig. 70). It is the normality of the parents that makes it *de novo*.

It is very important to distinguish, by examining the karyotypes of the parents, between the *de novo* translocation and the inherited variety. The former may be regarded as a chance mishap, comparable to the accident of non-disjunction. The recurrence risk in other children will be very low. The latter, as we shall see, has a high risk of recurrence within a sibship.

Inherited Translocation Down's Syndrome

It can happen, as we have pointed out in Chapter 5, that a person who

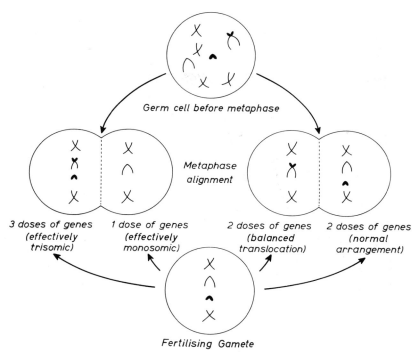

Fig. 71. *Karyotype of balanced (14;21) translocation carrier.*

seems, and indeed is, quite normal phenotypically can have one chromosome 21 translocated to another (Fig. 71). There is no excess, nor significant loss, of genetic material. This is a 'balanced' translocation.

But, let us restate, gametogenesis must present problems (Figs 41, 72).
Four types of gametes can be produced: those with both representatives
of pair 21, those with none, those with a single representative and 'free',
and those with a single representative of pair 21, but translocated as in
the parent.

Fertilisation of the gamete with two representatives will result in a
zygote with a translocation trisomy, in this case an 'inherited translocation Down's syndrome' (see Fig. 70). Fertilisation of the one with no
chromosome 21 will result in a monosomic zygote which, almost certainly, will come to nothing. Fertilisation of the gamete with a single
'free' chromosome 21 will result in a zygote that is normal both cytogenetically and phenotypically. Fertilisation with the single chromosome
translocated to another will result in a zygote with a balanced transloca-

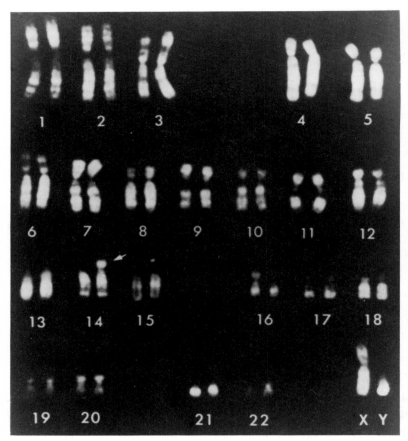

Fig. 72. *Gametogenesis in a balanced (14;21) translocation carrier and fertilisation of each of the four gamete possibilities.*

tion, normal phenotypically but destined in its turn to have the same problems of gametogenesis as its parent. Four zygotes: two with unbalanced complements, one with a balanced translocation, one entirely normal.

In theory, then, to such a translocation-carrier parent, one child in every four conceptions will have Down's syndrome. Since one in every four conceptions will be, almost for sure, a 'blighted zygote', one child out of every three born might have Down's syndrome.

In practice it does not work out that way. If the father is the balanced carrier the risk of a Down's syndrome child is quite low, say a 3 per cent

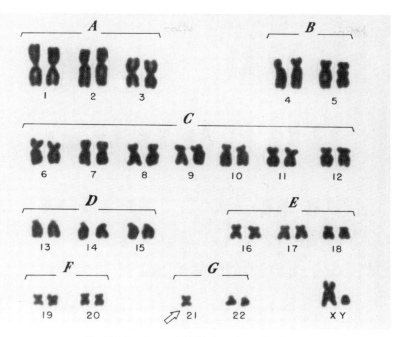

Fig. 73. Homologous 21/21 balanced translocation.

risk or 1:30. It seems that the abnormal sperm is disadvantaged, like a racehorse handicapped with extra weight. In addition there is, no doubt, discrimination against the abnormal zygote, for we know (see Fig. 49) that many Down's syndrome embryos are rejected as spontaneous abortions.

If the mother is the carrier the risk is nearer to theoretical expectation, but, even so, the risk is more like 10 per cent or 1:10 rather than 1:3. Such cases are few, four only in Uchida's 453, but they do happen and must be remembered.

There is worse to come. There are a few balanced translocation carriers in which there is a t(21;21) translocation. Both 21 chromosomes are joined together (Fig. 73). At gametogenesis they must move together. One gamete must have both of the pair, the other none. Zygotes must be either translocation trisomies (Fig. 74) or monosomies. A parent who has a t(21;21) translocation cannot have normal children.

Since the majority of Down's babies born to older mothers are due to

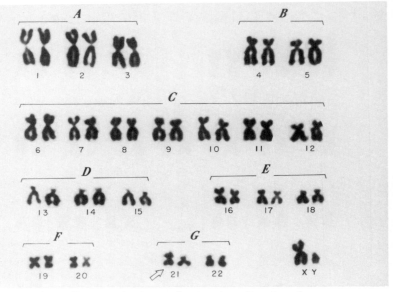

Fig. 74. Karyotype of 21/21 translocation trisomy Down's syndrome.

non-disjunction it is among the babies of young mothers that translocation mechanisms are more likely to be found. It is especially in families
where near relatives have Down's syndrome that the inherited translocation Down's syndrome will be uncovered.

Management of the Child with Down's Syndrome

The doctor who deals with the child with Down's syndrome faces more
than the problems of that child alone. The whole family needs help, for it
is upon the whole family that this tragedy has fallen: parents, sibs, grand-
parents—even more distant relatives. Much delicacy of rapport is
needed if the misfortune is to be mitigated and the poor parents guided
to accept in the best way possible the new and unthinkable situation. 'In
one short hour all their hopes, dreams and plans are shattered.' They will
suffer agonies of disbelief, anger, rejection of the bad news and often
anger at, and rejection of, the physician who has given them the bad
news. There may be guilt about something done, or left undone, in the
pregnancy that, in their minds, may have caused the disorder. To them
there has to be an explanation. They will be torn between a love that has
been 9 months in preparation and an unwillingness to accept an imper-

fect baby when their hopes and ambitions for this child have been so high.

There are few tasks more dismal than telling new parents that their new baby has Down's syndrome. What should the doctor do if he suspects that diagnosis in the first hours or day or two? Should he say nothing until time proclaims his apparent lack of diagnostic acumen or should he straightaway voice his suspicion? I believe he must be brave, but brave with caution.

The average practitioner will see few Down's syndrome babies and he may well be uncertain of the diagnosis. If he works in a centre, and most do not, where chromosome studies can be done he might arrange for a blood lymphocyte culture to be done immediately, asking the laboratory to give the matter high priority. He need say nothing at this point, unless asked by suspicious parents. If the diagnosis is proved in 3 or 4 days something certainly will have to be said. Or he may straightaway state his suspicions about Down's syndrome but say, with truth, that it is very difficult to be sure in a newborn baby—especially if it is premature and small—whether or not his suspicions are right. He can say that tests are in hand and that the result should be known in 3 or 4 days. Unless the baby is very big, the dermatoglyphics are difficult to read and it may not be possible to apply the nomogram (see Fig. 67).

If a specialist paediatrician who may have seen many Down's syndrome babies is at hand, let him be consulted, not necessarily with the parents' knowledge. The diagnosis may then be confirmed, refuted or still remain in doubt. If the paediatrician strongly confirms the practitioner's suspicions something will have to be said.

I have heard it taught that the father of the baby only, or some relative, should hear the bad news and that, for the moment, the knowledge should be kept from the mother. This surely is unrealistic. This would be a terrible secret to keep from the proud and unsuspecting mother of her newborn baby. No, both parents must be told, together, at the same time—not, one need hardly say, before a ward full of other mothers and their visitors!

However sure one may be of the clinical diagnosis, one must try to get cytogenic confirmation. Apart from the fact that grieving parents may question a mere opinion, it is important to know, not only that the diagnosis is beyond doubt, but to know also whether this is the 'regular' trisomic variety of Down's syndrome or whether it might have come about because one parent carries a balanced translocation. This author, in his younger days before chromosome studies had been invented, diagnosed Down's syndrome in, as it turned out, a normal newborn baby.

The repercussions of such an error are not easily forgotten. Seek confirmation, be sure, then tell the truth. One must explain that this is not a very rare disease, that one in every 750 babies born has Down's syndrome. It could—and does—happen to anyone. No one is immune from the possibility of a Down's syndrome baby. One must explain that, except in very rare circumstances, it is a chance mishap in which either an egg cell or a sperm cell has split unequally so that the factors of inheritance have become unbalanced. There is an excess, not a deficiency, of factors of inheritance: an imbalance of genes. Too many orders or instructions have led to confusion as to how the body will form, grow and mature. One must be sure that the parents fully realise that neither parent is to blame. No omissions nor commissions in pregnancy—or before—have been responsible. It is for that reason that I personally do not think it a good plan to try to determine by 'marker' studies (see Fig. 54) which parent contributed the extra chromosome. Even if the parents accepted that the contributor was in no way to blame, inlaws might be less understanding!

It is perhaps unwise to place too much emphasis on maternal age after the fact lest guilt might be felt at having a baby at an older age, or resentment might be engendered that there had been insufficient direction towards prenatal diagnosis. To risk guilt and recrimination at this point can do no good.

One must be frank and say that this baby, lovable though he will be, will be slow in development though that will not be obvious at first. He will be happy, cheerful and content but will not be able to benefit from ordinary school. To try to push him too hard and to have unrealistic expectations can only lead to mutual frustration and discouragement. One must applaud every achievement but not expect too much too quickly. Whether structured 'stimulation programmes' lead to greater achievements in the end than normal loving care and natural play is somewhat dubious. If it can be arranged, so much the better. The parent will feel that something positive is being done, and for that they will be most grateful. Nothing could be more cruel than for them to think that their baby and themselves have been 'written off' as beyond any help.

One must tell the parents to be frank and open with their relatives and friends. They must talk freely about their misfortune. This is not a matter for shame or concealment. Imagine the heartache of a mother who knows but conceals the facts, who tries to respond in cheerful terms to well-meaning inquiries about her new baby. If concealment is attempted the family will avoid their relatives and friends, at the very time when they are most needed. When the truth is out, there is no more to be said.

This author feels that the Down's syndrome baby deserves all that can be offered in the way of medical and surgical care. If the newborn baby has duodenal atresia surgery should not be denied to him if the parents wish that all that can be done should be done. If they adamantly refuse surgery on his behalf, the matter becomes most difficult indeed. It is not they who must stand by and watch the baby starve to death; it will be the nurses and doctors. On the other hand it is not the doctors and nurses who will have to rear a handicapped child who may have survived because of pressure put upon the parents to agree to operation. In some jurisdictions the law may rule that treatment cannot be withheld and that the parents' refusal to consent can be overruled. If at all possible it is much much better to avoid legal confrontations. If the baby has congenital heart disease which may require surgery the pros and cons of operation are usually less difficult to resolve than the arguments for or against newborn relief of duodenal atresia. By the age—often between 1 and 2 years—when cardiac surgery may be indicated, surgery will very rarely be refused.

The Down's syndrome baby will walk, but will be late to do so. He will talk but be slow and simple in his use of words. He will toilet train, but it may take a little longer. He may attend a day school for retarded children if one is near at hand; if not, he may have to go away to a residential school. In either event he will be happier among those of his own kind, not teased by normal children.

Unless he has congenital heart disease his health in general will be good, though he is likely to have more than his fair share of troublesome respiratory infections. And apart from these the child with Down's syndrome has two health problems. There is an increased likelihood of autoimmune thyroiditis, Hashimoto's disease, and its occasional sequel of hypothyroidism. This must be kept in mind. There is also a slightly increased risk of autoimmune pancreatic damage and diabetes mellitus and, as has been mentioned, a small but increased risk of leukaemia.

Many suffer from deprivation of social contacts and from loneliness. They are not accepted by the friends of their sibs. They are not so retarded that they lack such feelings. As adolescents they are not immune from sexual interests though usually their libido is low. Parents sometimes enquire about sterilisation of their Down's syndrome adolescent. They can at least be reassured that males are invariably infertile, the girls too have a very low fertility and, in any event, can be made to understand the consequences of promiscuity. In the society from which I write, it is not legal to sterilise an adolescent, however much retarded.

The Down's syndrome child will not become self-supporting and the parents cannot look after him for ever. They are likely to be elderly and perhaps by adult life he will be alone. For such reasons, most Down's syndrome patients come into some form of custodial care in time; usually nowadays it is in a 'group home' rather than in a large institution. Circumstances of course alter cases, but the usual advice is to rear the Down's syndrome child at home, if only for a few years. The parents will feel, at least, that they have done their best.

What of the other children? Will the Down's child in their midst harm them? If their neighbours and their children know the facts, if their playmates know the truth, if the parents feel no shame or guilt, no harm, I think, will come to the sibs of the Down's syndrome child. Knowledge of the facts will bring acceptance. Secrecy will only bring taunts and ridicule.

Is there any treatment? Unfortunately there is not. All kinds of therapies have been suggested. Special diets and megadoses of vitamins have been hailed as having great success. *There is no cure.* This must be said quite firmly. Desperate parents may otherwise take their child from doctor to doctor, from charlatan sometimes to charlatan, from continent to continent, even, seeking what does not exist. Disappointment only, and financial loss, will be their rewards. One must help the child to be the best he can be, accepting his limitations, not seek the unobtainable.

Genetic counselling is not concerned solely with risk facts and figures. Explanations as to the cause must be given. Guilts must be dispelled and recriminations between spouses and families dissipated. The genetic counsellor must concern himself with those things while remembering that it is not his function to be much directive, certainly not authoritarian, in the matter of having children—or more children. Only parents know how much they are prepared to risk to attain their hearts' desires. But what are the risk facts and figures? What are the chances of a couple having a Down's syndrome child—or indeed another Down's syndrome child.

A mother who has had no Down's syndrome child and none among close relatives has a risk related to her age: let us say one chance in 2000 at age 25, one in 800 at age 30, one in 350 at age 35, one in 100 at age 40, one in 25 at age 45. Figures vary somewhat from one study to another, but the above are not far out. Paternal age has little or no effect, as mentioned above.

To put these risks in perspective playing cards, 52 to a deck, are a useful illustration. At age 25 the risk is roughly drawing a single joker first draw from 40 decks of cards; at age 30 from 16 decks, at age 35 from seven decks, at age 37 from five whole decks and at 40 years of age a single joker from two decks. Even at age 45 the risk would be to draw the knave of hearts from the two red suits, hearts and diamonds. People understand cards better than percentages and odds!

When one must advise a mother who has already had a child with Down's syndrome, one must first know if at all possible the cytogenetic mechanism responsible. It will be recalled that it can have come about by the chance mishap of non-disjunction, as a *de novo* translocation, as an inherited translocation, by isochromosome formation or as a mosaicism.

If it has come about by non-disjunction, if it is the 'regular' trisomic variety of Down's syndrome, one is on somewhat insecure grounds for it is not certain whether some people, and this could apply to either partner of the couple, are at special liability to non-disjunction. One cannot be sure whether the couple that you are counselling might not be a high-risk couple. One has to rely on overall observed risks of recurrence for non-disjunction trisomy on empiric risk figures. Most people agree that, until the maternal age-related risk exceeds 1 per cent (by about 40 years of age) it is reasonable to give a 1 per cent recurrence risk; when the maternal age-related risk exceeds 1 per cent, that risk takes precedence. At age 35, when the age-related risk of occurrence is 0.3 per cent, the recurrence risk might be given as 1 per cent; at age 45, when the occurrence risk is about 4 per cent, that figure would be used to define the risk of having a second child with Down's syndrome.

If it is shown that the child has Down's syndrome as an unbalanced translocation trisomy, it is imperative that the chromosome complements of the parents be studied. One must know if they are quite normal or if one or the other carries a balanced translocation involving chromosome 21 (see Fig. 71).

If neither parent has a translocation, it is, in the child, a *de novo* translocation trisomy. Figures for recurrence of Down's syndrome in this situation are scanty. Rather too simplistically, I am sure, I am inclined to regard the *de novo* translocation trisomy as a special form of non-disjunction trisomy in which the extra chromosome has, irrelevantly, attached itself to another. I am inclined to give the same recurrence risk as I would give if the first Down's syndrome child had been of the 'regular' non-disjunction variety.

If one or other of the parents is shown to carry a balanced translocation, say a t(14;21) translocation (Fig. 71) one must take account of the problems of gametogenesis in such a person (see Figs 41, 72). As you will see the chances for the zygote, in theory at any rate, are: one Down's, one monosomy 21, one balanced carrier like the parent, one entirely normal. If one assumes, fairly in the vast majority of instances, that the monosomy comes to nothing, one could give, in theory at any rate, a one-in-three recurrence risk. It does not work out that way in practice for there is discrimination against the abnormal zygote and, apparently, additional discrimination against the abnormal sperm.

When the mother carries the translocation the risk of another child with Down's syndrome is not around 30 per cent; it is no more than about one in ten; or to be more optimistic, ten to one against that misfortune. If the father carries the balanced translocation the double-discrimination reduces the recurrence risk far below theoretical expectation: no more perhaps than a 3 per cent risk, or even less.

When one encounters such a hereditary translocation in a family one must think how far and wide one's responsibilities extend. The translocation carriers themselves acquired the translocation—but from where? To be sure, they may have acquired it *de novo*, or they may have acquired it from a parent (who may not have had a Down's syndrome child as evidence of that possession). It could be, then, that the sibs, nieces and nephews and cousins of a person might be at significant risks also of having children with Down's syndrome by inherited translocation. My practice is to set out in writing who among the relatives might be at risk, and to give a copy of this exposition to the consultand. He or she can decide how much to tell to whom. He or she knows better than myself how much, or how little, her relatives would like to receive such warnings. I do not think that I myself, directly, should inform the relevant relatives concerning what I have found out from a study of my patient.

If the Down's syndrome child is shown to have an homologous t(21;21) translocation (see Fig. 74), this may have come about *de novo*, or it could be (but very rarely) that one or other parent has a balanced t(21;21) translocation (see Fig. 73). If such were to be the case, such a parent could not have a normal child at all. Both 21 chromosomes must be in one gamete; the other has none. A zygote must have either a translocation trisomy or must be monosomic for chromosome 21. It can happen.

Down's syndrome by isochromosome of the long arms of chromosome 21 (see Fig. 43) is a very rare and chance event. The recurrence risk can be stated as very low indeed, probably no greater than the age-related risk.

Down's syndrome as a mosaicism is an event unique to that individual. The two chromosome complements in two stemlines of cells gives evidence of non-disjunction, not in the parents, but in the zygote: a chance error of mitotic non-disjunction. As such the risk of recurrence in a sib is minimal, perhaps non-existent.

Prenatal Diagnosis in Down's Syndrome

All the risks given above are, of course, for a liveborn baby with Down's syndrome in the absence of any modifying interferences. In fact one can say to a consultand, whatever their risk for whatever reason, 'we can nearly guarantee that you will not have a liveborn baby (or another baby) with Down's syndrome provided you accept the idea of prenatal diagnosis, with the further acceptance (and this should not be a condition to the offer of prenatal diagnosis) of the idea of termination of a proved abnormal pregnancy'.

The techniques, advantages and disadvantages of prenatal diagnosis by amniocentesis or by chorion villus sampling have been discussed in general terms in Chapter 3; but what can one say related specifically to Down's syndrome?

Any couple of any age are at risk of having a baby with Down's syndrome, though at a maternal age of 25 it may be no greater than the chance of drawing a single joker from 40 decks of cards. Any couple might wish prenatal diagnosis. Apart from the very slight (in the case of chorion villus sampling, uncertain) risk of pregnancy loss due to the procedure, even in the affluent society from which I write, constraints of cost and manpower make universal prenatal diagnosis impracticable. Most clinics offer prenatal diagnosis to women over 35 years of age though one admits the illogicality of drawing a firm line between age 35 and 34.5. One would not refuse amniocentesis at any age to a woman distraught by anxiety—perhaps because her best friend had had a mongol baby.

One certainly would offer, indeed encourage, prenatal diagnosis if one or other parent were known to be a carrier of a balanced translocation. If the complement of chromosomes of the fetus is entirely normal, well and good. If there should be a balanced translocation, also well and good. The child will be as normal as the parents—though it will have the same reproductive problems as them in years to come. It is unlikely that at amniocentesis one will encounter a monosomic fetus. It almost certainly will have been a blighted zygote or rejected as a miscarriage before the time has come for amniotic tap. One might, though I have seen no such report to date, discover a monosomy 21 at chorion villus sampling. Should such be found, termination of the pregnancy would be offered.

If one should discover a fetus with a 21 trisomy one should again go over with the parents the implications of that finding and recount in detail the features and future of the child with Down's syndrome. The decision to continue or to terminate the pregnancy must be theirs and theirs alone. There must be no pressures nor influence one way or the other. Whatever they decide is right for them and they must be supported in that decision.

One cannot make a strong case on recurrence risks alone for prenatal diagnosis in a pregnancy following the birth of a child with Down's syndrome due to *de novo* translocation, isochromosome formation or mosaicism. The risk of recurrence is very low, but it may well be worthwhile for the parents to have the reassurance of a normal fetal chromosome analysis. Good news is always welcome.

Prenatal diagnosis is a happy occupation. Very rarely does one give bad and tragic news. In the great, great majority of cases the news is good. Even in the mother of 40 years of age, the risk of a Down's syndrome child is no more than her chance of drawing a single joker from two decks of cards.

One looks forward to the day when chorion villus sampling will be routine and allow, if necessary, much earlier abortion. Abortion at 20 weeks, the usual time after an amniocentesis diagnosis, is a miserable, distressing business.

CHAPTER 8

Other Autosomal Abnormalities

One might imagine that since such a stereotyped abnormality as Down's syndrome is invariably associated with a chromosome anomaly, other comparable associations might have been revealed. One might have expected that cases of Cornelia de Lange's syndrome, the Smith–Lemli– Opitz syndrome, the Meckel and the Moon–Biedl syndromes might show major chromosome abnormalities, but such is not the case. Only quite rarely is a stereotyped, regularly recognised syndrome associated with a regularly identifiable chromosome abnormality. There have been, to be sure, some gains in the last few years but they have not, with perhaps one exception, been spectacular.

It seems that very few losses or gains of whole autosomal chromosomes are compatible with intrauterine life. The aneuploidies of almost all the autosomes come to nothing; they are either very early blighted zygotes or early miscarriages. Trisomy of chromosome 16 is rather common in miscarried pregnancies, but it is never seen in a liveborn neonate. With few exceptions fetuses with autosomal trisomies and monosomies do not live long enough to be born alive as 'syndromes'.

Bits and pieces of chromosomes missing or in excess, the partial deletions and partial additions, do not result very often in stereotypes. Two bits may look the same by the crude tool of microscopy, but they may in function be very different. Two patients may both appear to have, let us say, a similar deletion from the long arm of chromosome 7. One might expect them to have the same clinical features, the same phenotype. But they do not. The missing pieces may look alike but they may not be *genetically* identical. There are dozens, maybe hundreds, of gene loci to a chromosome. One partial deletion might cause the loss of most important genes that are not lost by what appears—but only appears—to be a similar deletion. Even the most sophisticated visual cytogenetics is a crude tool indeed.

This author finds 'syndromology' and 'dysmorphism' difficult, frustrating and humiliating. One should be able to say, with some confidence: 'This baby will be found to have a partial deletion of the long arm of chromosome 7' or 'that baby has a trisomy of part of the short arm of

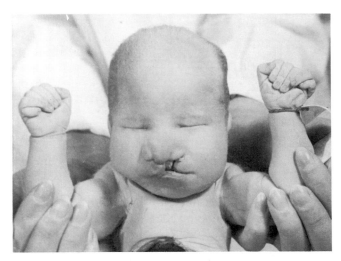

Fig. 75. Phenotype of baby with trisomy 13.

chromosome 4'. But one cannot; at least this author cannot. New tools are in the making. DNA 'markers' are now at the point of development that it is possible to 'walk the chromosome' identifying the gene loci one by one in order along the way. The recognition of genes missing or in excess, rather than additions or deletions of bits and pieces of chromosomes, may make syndromology more rational.

Certain features, presumably because of especial vulnerability of embryonic tissues, tend to be seen in the chromosome disorders. Except in the sex chromosome disorders mental retardation is invariable when bits and pieces—or whole chromosomes—are missing or in excess. Low birth weight for gestational age is very common. Micrognathia, cleft lip and palate, malformed and low-set ears, abnormal dermatoglyphic patterns, abnormal genitalia in the male and congenital heart disease are very often seen in the chromosome disorders.

The 13-Trisomy Syndrome (Patau's Syndrome)

This syndrome, first recognized by Bartholin in 1657 and redefined by Patau in 1960, is a rather rare abnormality, perhaps seven times less com-

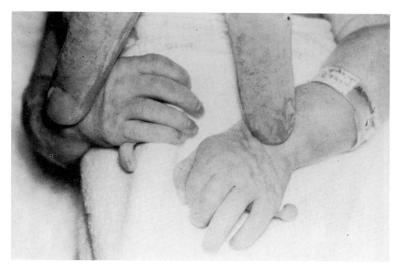

Fig. 76. Polydactyly in trisomy 13.

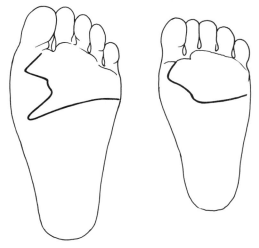

Fig. 77. Fibular arches on hallucal areas in trisomy 13.

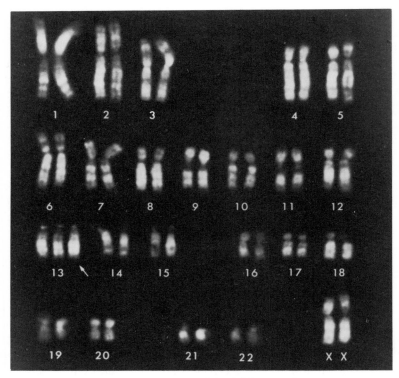

Fig. 78. Karyotype of trisomy 13; Q-banded fluorescent karyotype.

mon than Down's syndrome. Formerly it was called the D-trisomy syndrome until banding staining made possible the individual recognition of chromosomes within group D.

Most babies are of low birth weight for gestational age, frail, puny and microcephalic. It is the forebrain that is especially underdeveloped with special dysgenesis of the olfactory lobes—arhinencephaly. There may be complete failure of development of the forebrain—holoprosencephaly. Microphthalmos is almost universal, and there may be almost complete anophthalmia. Occasionally the eyes are fused together centrally—cyclopia. Almost always there is cleft lip and palate. The ears are small, ill formed and low-set. Diffuse haemangiomas are common. The neck may be partly webbed with loose redundant skin (Fig. 75).

There is very commonly polydactyly of the hands (Fig. 76), less commonly of the feet. The dermatoglyphic patterns are abnormal. The palmar axis triradius is almost always displaced distally to the t" position (see Fig. 65). The usual digital patterns are arches. Transverse palmar, simian creases (see Fig. 68) are the rule, but the hallucal pattern is the most characteristic of all. In this location there may be a fibular arch (Fig. 77). Such a pattern made the diagnosis in the baby shown in Fig. 75 even though it had been preserved as a museum specimen for 30 years! This author has seen a baby who while lacking many of the features of the syndrome, had such a fibular arch. A chromosome study confirmed trisomy 13.

Almost all these babies have congenital heart disease of one kind or another, usually a ventricular septal defect. Many have kidney abnormalities which may include polycystic kidneys. In the male external genital malformations are common. Females often have bicornuate uteruses.

These babies live just a few days or weeks, occasionally a year or two, and severe mental retardation is the rule.

Trisomy of chromosome 13 is the essential cytogenetic feature. Usually it is 'free' as in Fig. 78, but it may be attached as a translocation trisomy, almost always to another D group large acrocentric chromosome. Such a translocation may come about *de novo* or as inherited translocation from a carrier parent. As in Down's syndrome the risk of a carrier parent having a child with a translocation trisomy of chromosome 13 is, in theory, very high indeed. In fact there is very great discrimination against the abnormal zygote and, even with a translocation carrier parent, the risk of recurrence is very low indeed. Prenatal diagnosis by amniocentesis or chorion villus sampling can and should be offered.

There is an important differential diagnosis to the trisomy 13 syndrome: Meckel's syndrome. The clinical features can be very similar: microcephaly, microphthalmia, cleft lip, polydactyly, congenital heart disease and polycystic kidneys. True, the baby with Meckel's syndrome usually has an occipital encephalocele, but not always. The chromosome complement is normal. If a baby looks as though it has trisomy 13, but has a normal chromosome complement, it is not a case of the trisomy 13 syndrome; it must be regarded as having Meckel's syndrome.

Does this matter? It does, because Meckel's syndrome is by autosomal recessive inheritance with a 25 per cent recurrence risk—and there is no means of prenatal diagnosis.

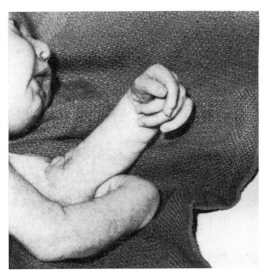

Fig. 79. The hands with tightly clenched fists in a baby with trisomy of chromosome 18.

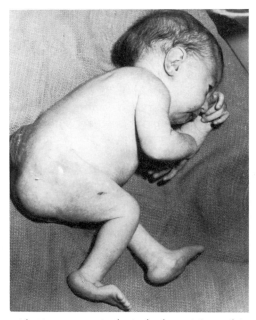

Fig. 80. Baby with trisomy 18; note the 'rocker-bottom' feet and (in this case) the meningomyelocele.

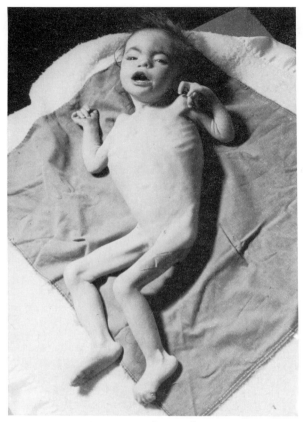

Fig. 81. Trisomy 18 with survival to age 5 years.

The 18 Trisomy Syndrome (Edward's Syndrome)

This is not so very rare; it is more common than trisomy 13. The incidence might be as high as 1:3000 births. The pregnancy is almost always full term, often prolonged. Female babies are four times more commonly affected (or survive to be born alive and recognised for what they are) than males. Again there is a relationship to maternal age.

The occiput is prominent, the chin receding; there is microcephaly. The ears are low-set, flat and elf-life, the nose is pointed and 'beaky'. Congenital heart disease is common; cryptorchidism is constant. Inguinal, lumbar and umbilical hernias are often present, and the pelvis is characteristically very narrow. There may rather commonly be a lumbar

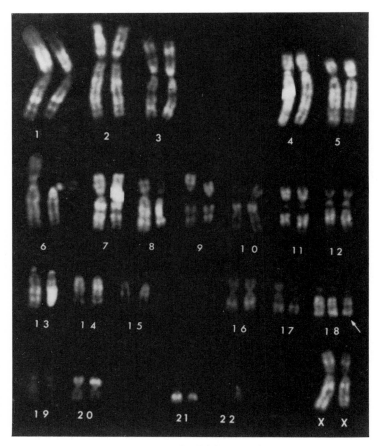

Fig. 82. Trisomy 18; Q-banded fluorescent karyotype.

meningomyelocele. Variable defects, or no visible abnormality, are found in the brain.

The most striking feature of these babies is the rigidity, especially of the arms. This does not seem to be a central nervous system spastic cerebral palsy but rather a joint and ligamentous fixation. The arms can scarcely be fully extended and the fingers are rigidly flexed across the palms with the index finger often overlapping the middle finger (Fig. 79). Simian palmar creases are usual; digital dermatoglyphics are almost all arches. The feet are characteristically 'rocker bottom', like the rockers of a rocking-chair (Fig. 80).

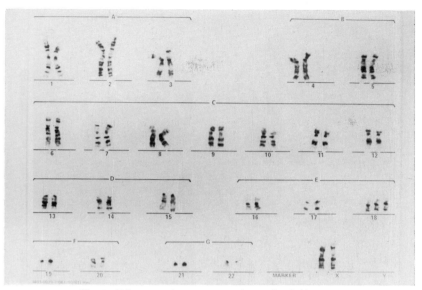

Fig. 83. Trisomy 18; for comparison, Giemsa-banded karyotype.

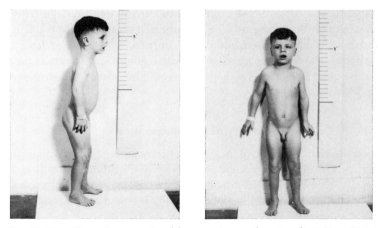

Fig. 84. Gary P, age 3 years, referred because of unusual posture, limitation of joint mobility and unusual facial appearance.

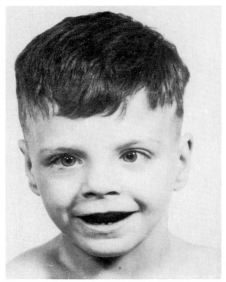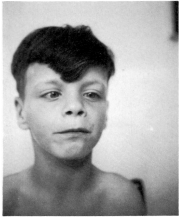

Fig. 85. Gary P: facial appearance aged 3 and 10 years.

These babies, of low birth weight despite a long gestation, are quite small and frail. Usually they die in a few days or weeks, but they can live for years (Fig. 81).

The essential chromosome abnormality is trisomy of chromosome 18. Usually it is 'free' (Figs 82, 83), but as with Down's syndrome or trisomy 13 the extra chromosome may be translocated to another, and this may arise *de novo* or be inherited. Some cases of trisomy 18 are mosaics.

Although the recurrence risk, even with a translocation carrier parent would be quite low, prenatal diagnosis would be offered following a trisomy 18 child.

Trisomy of Chromosome 8

The only other trisomy of a whole chromosome that is other than extremely rare in a living child is trisomy of chromosome 8. It is less frequently recognised than trisomy 13 or trisomy 18 but is probably quite often missed, especially since it usually is found as a mosaicism.

Case History

Gary P was referred at 3 years of age by the orthopaedic surgeons because

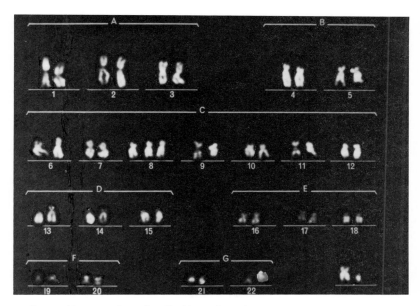

Fig. 86. Karyotype from skin fibroblast culture showing trisomy of chromosome 8; chromosomes in blood lymphocytes were normal.

of his unusual posture, limitation of joint movements and camptodactyly (Fig. 84). He had an unusual facial expression (Fig. 85). Blood lymphocyte chromosome analysis was normal, but because of his unusual features and very many digital arch patterns, a skin fibroblast culture was done. These cells showed a universal trisomy 8 chromosome complement (Fig. 86).

Gary has now grown up. He is a pleasant young man (Fig. 87). He has a slight sensineural hearing deficit and slight mental retardation. He is able to sustain conversation, and do simple work. His ambition is to save money to buy a powerful motorcycle! His persisting camptodactyly (Fig. 88) is no handicap.

If Gary were to have children—and he might—I do not know what the outcome might be. I do not know if the trisomy might be present in his spermatocytes, as in his skin fibroblasts, or absent as in his blood lymphocytes.

The facial appearance of trisomy 8 seems characteristic and Gary is a good example. As in Gary, osteoarticular lesions with limitation of joint movements are common. Camptodactyly is usual. A characteristic feature, not remarkable in Gary, is great thickening and furrowing of the skin of the palms and soles. Gary's mental status is about average for trisomy 8. Mild mental retardation is the rule, and severe intellectual deficit

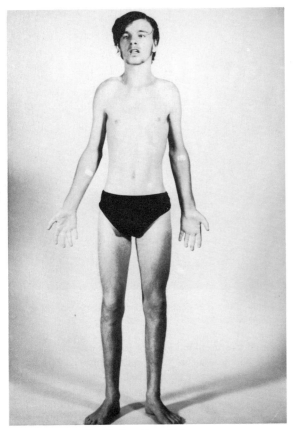

Fig. 87. Gary P aged 20 years.

is unusual, but I have seen it in another adult patient. Some are not retarded.

Trisomies of chromosome 9 and 14 have been reported, but less than ten cases of each are on record in the literature. Trisomy of chromosome 22, at one time described as a distinct entity is not now thought to be compatible with life. Our patient, Susan (see Fig. 97, page 150) was at one time thought to represent trisomy 22. Banding patterns now show her to have partial trisomy of chromosome 14 (see Fig. 98, page 151).

Partial Trisomies and Deletions

For reasons that have been explained, partial deletions and excesses

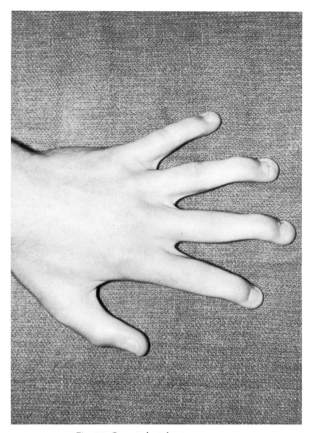

Fig. 88. Camptodactyly at age 20 years.

rarely produce a stereotyped syndrome. Some do, most do not. There is almost no limit to the syndromes, or imagined syndromes, that have been described as associated with this or that less than complete chromosome anomaly. Each one of us working in the field of clinical cytogenetics has acquired his personal collection of chromosomal errors and clinical oddities. The second edition of de Grouchy and Turleau's *Clinical Atlas of Human Chromosomes* (1984) is a magnificent compilation of almost all the reported chromosome anomalies and their clinical counterparts. Here in this book will be described only a few of the more stereotyped syndromes and some examples from this author's own collection of cytogenetic treasures.

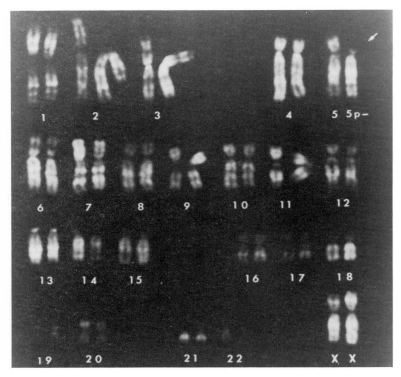

Fig. 89. *Deletion of the short arm of chromosome 5: the cause of the cri-du-chat syndrome.*

Cri-du-Chat (Cat-Cry) Syndrome, 5p —

This syndrome, regularly associated with deletion of the short arm of chromosome 5 (Fig. 89) is not rare; it can be recognized with some degree of certainty.

These babies, who may survive for many years, exhibit, at least in early infancy, a clearcut picture. There is a round moon-face (Fig. 90), microcephaly, hypertelorism and low birth weight. There may be epicanthic folds and micrognathia. The dermal patterns are not distinctive though transverse palmar creases and a distal axis triradius are common.

It is the cry that is striking. It is quite strange, a plaintive, high-pitched wail, weak with a hint of stridor. It really does sound like the mewing of a kitten. At one time when we had such a baby in our nurseries a cleaning woman spent some time searching for the stray kitten. This cry re-

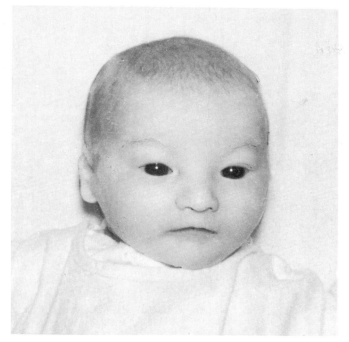

Fig. 90. Jamie G as a newborn baby with the cri-du-chat syndrome.

sults from laryngomalacia, with a narrow glottic opening and a curved epiglottis. On pronation there is an air leak at the posterior commissure; the strange cry comes from approximation of the cords in front. As the babies grow older this feature disappears. While they tend to develop stridor easily with upper respiratory infections the cry is no longer distinctive after a few months.

While the facial appearance changes and is no longer rounded it is, from personal experience and published pictures, still rather characteristic (Fig. 91). A correspondent, thousands of miles away, sent me a photograph of their son with the cri-du-chat syndrome. Their child and my patient are as alike as two peas; they could be monozygotic twins.

There is perhaps a distinctive feature present in two out of three cases seen by myself, and in others illustrated in the literature: a flat, sessile nodule, 2 or 3 mm in diameter a centimetre or so in front of each ear.

The chromosome disorder is a deletion of part of the short arm of chromosome 5. More specifically it appears that it is deletion of the bands 5p14 and p15 that is responsible for the syndrome.

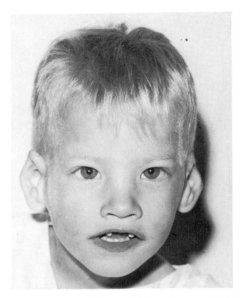

Fig. 91. Jamie G age 7 years, profoundly retarded but with no peculiarities of voice at this age.

Most cases arise *de novo* as an error at gametogenesis but some cases do arise from a parental balanced translocation. In the parent 5p may be translocated elsewhere in the chromosome complement. The parent will not lack 5p; it will be there. The gamete (and thus the zygote) may well receive chromosome 5 with the deletion, but may not receive the chromosome to which 5p is translocated.

One case of mine, Nancy M, has typical cri-du-chat. She shows the 5p − deletion. Her father, quite normal, shows a long and very narrow constriction—a fragile site—on one 5p. Part of the short arm is carried, as it were, on a stalk. At gametogenesis the stalk evidently has broken and the distal fragment has become lost.

Chromosome 4, Short Arm Deletion, 4p −

Many cases of this quite distinctive syndrome have been described under the fanciful name of the 'Greek warrior syndrome'.

There is microcephaly but a high forehead. The nose bridge is wide and prominent and the nose wide, straight and with angular sides. The

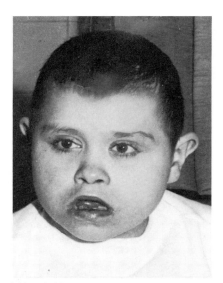

Fig. 92. Child with deletion of part of the long arm of chromosome 18.

result is the facial appearance resembling a Greek soldier's helmet. Cleft palate, colobomas and other facial dysmorphic features are seen. Half these babies have congenital heart disease. All are very severely retarded.

The Prader–Willi Syndrome

Until quite recently no cytogenetic abnormality was associated with this well-known syndrome. These children start as very hypotonic, frail and puny babies with feeding difficulties and failure to thrive. There is nothing very distinctive at first. At a year or so of age, while remaining hypotonic, they develop voracious and uncontrollable appetites. They will stop at nothing to get at food. They become very obese and may develop diabetes. There is mild to moderate mental retardation.

Just in the last 3 or 4 years it has been recognised that many of these patients—but not all—have a very small interstitial deletion, bands q11 q13, very close to the centromere in the long arm of chromosome 15. That this is not a universal finding suggests that the Prader–Willi syndrome is a syndrome only, a constellation of phenotypic features: no more than that. There may well be more than one cause, and more than one type, of Prader–Willi syndrome.

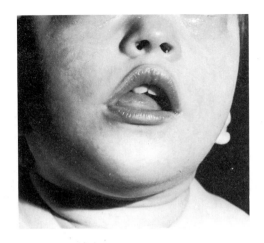

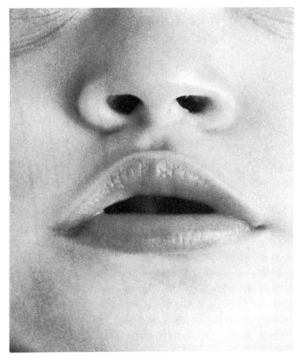

Fig. 93. The 'carp mouth' associated with deletion of the long arm of chromosome 18.

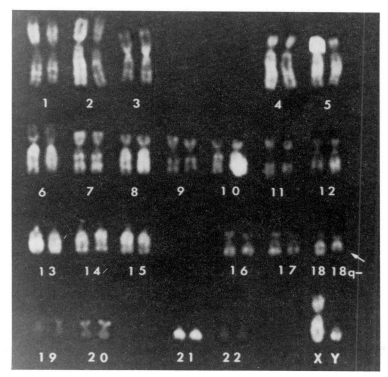

Fig. 94. Karyotype showing a major deletion of the short arm of chromosome 18.

Chromosome 18 Deletions

Deletions involving both short and long arms of this syndrome are not rare in living children. Perhaps chromosome 18 is less important than some. As we have seen trisomy of chromosome 18 is one of the very few autosomal trisomies compatible with life. Deletions, while serious, can likewise be survived, and for many years.

There is nothing very specific about the facial dysmorphism associated with 18 p −. The face is round, the nasal bridge is flat, the upper lip juts forward, the philtrum (the upper lip ridges) is flat and there may be no upper lip 'cupid's bow'. The ears are large, low-set and floppy. There is mild to moderate mental retardation and a normal life expectancy.

Deletions involving the long arm, 18q − are likewise compatible with a normal lifespan with very variable retardation. The phenotype is per-

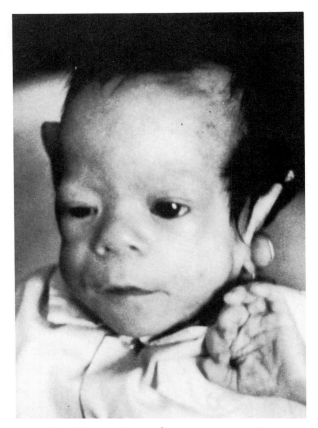

Fig. 95. Jeremy A, unusual appearance as a neonate.

haps more distinctive than in the 18p − syndrome (Fig. 92). The mouth, the so-called 'carp mouth' or 'fish mouth' is rather striking (Figs 92 and 93). The hands are small, the thumbs set back on the hand towards the wrist, and the fingers have an unusually larger number of whorl patterns. Again, as almost always with any but the sex chromosome anomalies, there is microcephaly and mental retardation. A karyotype is shown in Fig. 94.

An Interstitial Deletion of Chromosome 7, 7q −: Case History

Jeremy certainly was an odd-looking baby (Fig. 95) but there was not too much concern until, at a few months of age, there was obvious deve-

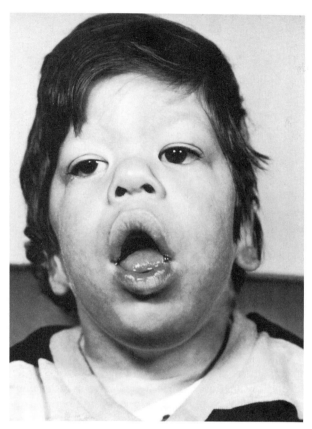

Fig. 96. Jeremy A aged 5 years when he was recognised as having an interstitial deletion in the long arm of chromosome 7.

lopmental delay. Assessment at 9 months revealed nothing specific. A chromosome study appeared normal.

As he grew older his appearance became more abnormal (Fig. 96). The parents, wishing to have other children, became more demanding for a factual aetiological diagnosis on which a recurrence risk for other children might be based. Another fluorescent, q-banded karyotype was made and most carefully scrutinised. This time a very small—and *de novo*—interstitial deletion in the long arm of chromosome 7 was discovered: 7q1 − : a partial monosomy in chromosome 7.

The lessons from this case? One may have to look extremely carefully for tiny abnormalities; some deletions or additions may be beyond

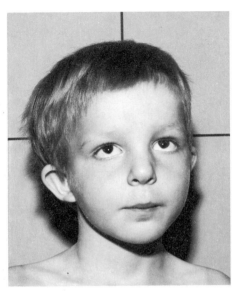

Fig. 97. Susan S: unusual appearance, mild mental retardation and congenital heart defect; at first was believed to have trisomy of chromosome 22 but later shown to have trisomy of part of chromosome 14.

microscopic resolution. Many dysmorphic syndromes may be classed as 'idiopathic' and of 'unknown cause' when they may be due to a small chromosome abnormality, but an abnormality involving many genes.

Partial Trisomy of Chromosome 14: Case History

Susan was recognised as looking unusual right from birth. She also had cogenital heart disease, a large atrial septal defect (Fig. 97).

Chromosome studies done many years ago showed a 47,XX complement. There was a small, apparently acrocentric chromosome in excess. At that time, it was thought to be chromosome 22. Susan was thought to be an example of a syndrome related to trisomy 22.

Years later the chromosome study was repeated with fluorescent staining. It was now clear that Susan has a partial trisomy of chromosome 14. De Grouchy and Turleau describing this happening say: 'It is characterized by a 47-chromosome karyotype, the supernumerary chromosome corresponding to the short arm, the centromere and the proximal portion of 14q'. That well describes Susan's karyotype (Fig. 98). De

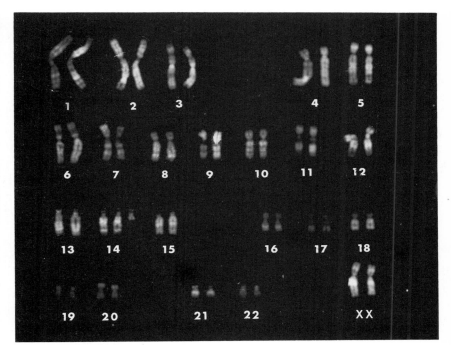

Fig. 98. *Karyotype of Susan S showing partial trisomy of chromosome 14.*

Grouchy continues: 'The length of the trisomic segment is variable. The site of the breakpoint varies from q12 to q24.' With such variability, and thus the possibility of great genetic variability, it is not surprising that Susan does not very closely resemble the several children illustrated by de Grouchy; nor do they much resemble one another.

The lessons from this case? Examples of what were formerly described as trisomy 22 are very probably a partial trisomy of some other chromosome. Cytogenetically similar abnormalities should not be expected to have closely similar phenotypes. Cytogenetics is a crude tool.

Translocation Partial Trisomy of Chromosome 13:13q Trisomy: Case History

David, too, was a most unusual-looking baby with a very narrow forehead (trignocephaly), a prominent bridge to his nose (glabella), a broad tip to his nose and anteverted nostrils. The maxillary alveolar arch was

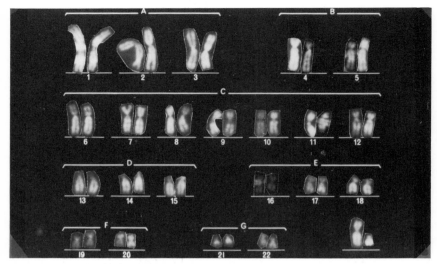

Fig. 99. Q-banded karyotype of David C; it was passed as normal, though in the light of later knowledge an abnormality can be seen.

very broad. Several fingers had only single creases. There was a right-sided simian crease. At this age the dermatoglyphics could not be read with certainty, and a chromosome study by fluorescent banding was reported as normal (Fig. 99). At 4 weeks of age he developed pyloric stenosis which was treated surgically.

The diagnosis was in doubt, but I thought he had some features of the autosomal recessive Smith–Lemli–Opitz syndrome. That at least was a working diagnosis, but it was not very satisfying, especially when as he grew older one found that his dermatoglyphics did not have a large number of whorl patterns, a feature of the Smith–Lemli–Opitz syndrome.

As he grew older, not only did the diagnosis become more dubious but it became more important, for the parents wished to have other children. I asked Dr Jung, a new colleague, to see David, now age 2 years (Fig. 100). He suggested that the chromosomes should be looked at again, this time by Giemsa banding. Now it was quite clear that there was an addition to the short arm of chromosome 3 (Fig. 101). What did this represent? One could not tell from this study only. The chromosomes of both parents and the sib to David were examined. The mother, and David's brother, both showed a balanced translocation. In both part of the long arm of chromosome 13 was translocated to the short arm of

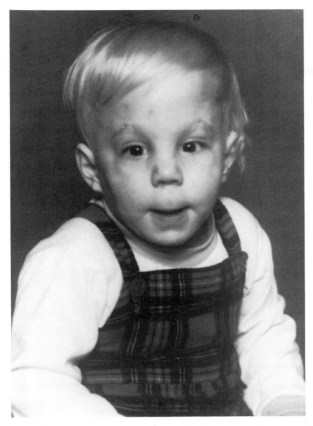

Fig. 100. David C age 2 years; at this time another chromosome study was done using Giemsa staining.

chromosome 3 (Fig. 102). David, by the mechanism set out in Fig. 42, had acquired the chromosome 3 with the addition, but he had also a pair of 13 chromosomes, complete and entire: he has a translocation trisomy of 13q involving the regions q2q3. He certainly resembles illustration 13.12 in de Grouchy and Turleau's *Atlas* which depicts a case of 13q2q3 trisomy. The mother's mother had a normal karyotype. The mother's father, an only child, had died. David's mother either acquired the translocation *de novo* or from her deceased father; we have no way of knowing.

The lessons from this case? Two heads are better than one; another opinion can set one straight. An alternative staining method may reveal

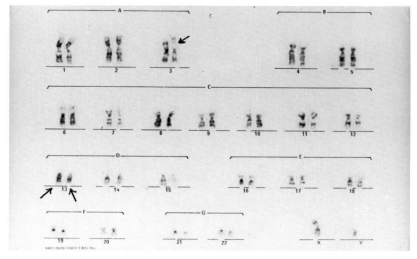

Fig. 101. Giemsa-banded karyotype showing an addition to the short arm of chromosome 3; note that the long arms of both 13 chromosomes are normal.

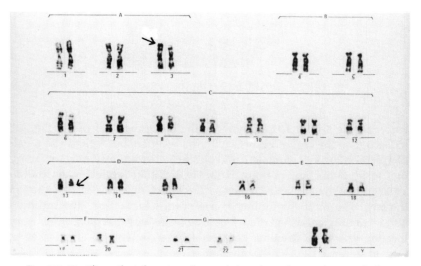

Fig. 102. David's mother's karyotype shows a balanced translocation between the long arm of chromosome 13 and the short arm of chromosome 3.

what had been passed up. Do not just write an order 'chromosomes, please'. Explain that the case does not correspond with loss or gain of a whole chromosome. It may be that a bit or piece is missing or in excess. Explain to the laboratory what they might be looking for. They will look better and more successfully that way.

Innumerable examples from the literature, and from personal experience could be cited to illustrate partial trisomies or partial monosomies, but perhaps the foregoing will suffice to make some points.

Recommended Further Reading

de Grouchy, Jean and Turleau, Catherine (1984). *Clinical Atlas of Human Chromosomes, 2nd Edition*, John Wiley and Sons, New York and Chichester.

CHAPTER 9

Chromosomes and Cancer

As a paediatrician I had long been aware of a group of childhood diseases in which there is a predisposition to develop cancers, leukaemia in particular. There is the Fanconi's anaemia syndrome (quite a separate entity from Fanconi's renal tabular dysfunction) in which there is bone marrow hypoplasia, growth failure, hypoplasia or aplasia of the radii and thumbs and a high risk of developing leukaemia in later life. There is ataxia telangiectasia with progressive incoordination, peculiar knots and leashes of blood vessels, especially noticeable in the conjunctivae (Fig. 103) and often immunological deficiency. Apart from Down's syndrome there are some other cancer-liable diseases. What do they have in common?

There is 'instability' of DNA because the enzymes responsible for repair of DNA (which is in all of us at all times in a constant flux of damage

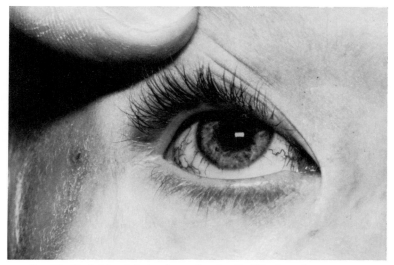

Fig. 103. Ataxia telangiectasia; abnormal conjunctival blood vessels at age 7 years.

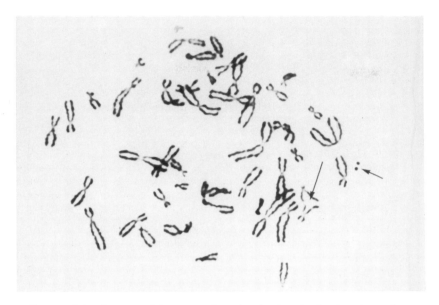

Fig. 104. *Metaphase spread from case of ataxia telangiectasia showing abnormal chromosomes.*

and repair) are defective. The instability of the DNA results in the chromosome breaks and rearrangements that are a feature of this group of diseases (Figs 104, 105) which are sometimes referred to as the 'chromosomal breakage disorders'.

It has also been recognised for a long time that cancer-producing agents, ionising radiation and chemical carcinogens, are usually mutagens: they cause genetic mutations and chromosomal rearrangements. There are also some families in which certain cancers seem to have a hereditary component in their aetiology. Some retinoblastoma eye neoplasms seem to behave as though determined by dominant inheritance, albeit with incomplete penetrance.

Just these simple facts lead one to wonder if the statement 'Cancer is essentially a genetic disease . . . ', made in the *Lancet* review article cited below may be true indeed. Let us look at what recent research has revealed.

At the turn of this century, when chromosomes had been identified as the repository of genetic material, Theodor Boveri speculated on a chromosome imbalance in cancer. In 1910, Peyton Rous, then a young man

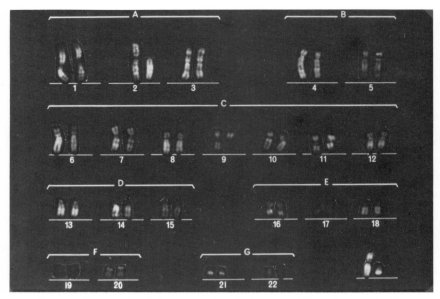

Fig. 105. Abnormalities of chromosomes 2 and 9 in a case of Fanconi's anaemia. The apparent absence of chromosomes 17, 19 and 22 is a photographic artefact.

working at the Rockefeller Institute for Medical Research, found that a cell-free filtrate of a chicken malignant tumour (which came to be known as the Rous sarcoma) could cause sarcomas in chickens. There appeared to be a tumour-producing virus. There was general scepticism of this work and it was abandoned until several decades later when the vindication of his work rewarded Rous with a Nobel prize at the age of 85! What has been the vindication?

Work with certain retroviruses, viruses that must be transformed by the enzyme reverse transcriptase before they can be incorporated into, 'infect' if you will, the DNA of the cell nucleus, showed that these viruses could carry genes that could produce cancers in animals: among them the Rous chicken sarcoma. These retroviruses could carry cancer genes: oncogenes. The virus oncogenes might be designated 'v-oncs'.

When methods used to identify these v-oncs were applied to human tissues it was found that normal human cells in all tissues of all of us carry oncogenes very similar, probably identical with, the v-oncs. These cellular oncogenes can be designated c-oncs.

Review of all animal species from yeasts to man have shown that through eons of evolutions these oncogenes have been preserved from phylogenetic generation to generation. This conservation implies importance. These oncogenes must be essential to the cells of the animals that they compose. It seems, moreover, probable that the v-oncs have been 'hitchhikers' on the viruses. When a virus invades a cell and multiplies therein it can pick up and incorporate into its own DNA or RNA the normal c-onc sequences of the DNA of the host.

What use are these c-oncs? We all know that the embryo develops from the zygote and the fetus from the embryo by cell proliferation and differentiation, and we know, furthermore, that the way the body forms and functions is determined by genetic controls, switched on or off at some points in time, switched on in some organs, not in others. It is not a big step to imagine that the c-oncs are the genetic controls of normal cell proliferation and normal cell differentiation. The c-oncs may be the essential genetic endowment of every cell. But it is not a great step further to imagine that enough is enough. That one can have too much of even a good thing. Perhaps overactivity of the c-oncs could cause too much proliferation; perhaps defective functioning of the c-oncs might cause failure of differentiation. A cancer is a wild and uninhibited proliferation of undifferentiated cells, perhaps due to wild activity of an over-enthusiastic c-onc.

The over-exuberance of a c-onc can be tested *in vitro*. If DNA is extracted from tumour cells and added to a culture of mouse fibroblasts growing as they normally do in a thin layer, a solid tumour of these murine fibroblasts forms in the tissue culture. We have a test for c-oncs that have become carcinogenic. But why might, or how might, a beneficient c-onc become a malignant influence? What might turn on a c-onc to destructively enthusiastic activity or alter its genetic code to change its instructions to the cell.

It might be that a virus invasion of a cell might turn on a c-onc, that the viral DNA or RNA might act as a 'promoter' (see Fig. 10). This does not seem to be an important mechanism in man.

It could be that a 'point' (single base-pair sequence change) mutation might so alter the c-onc that it codes for an abnormal protein product that in its turn causes abnormal proliferation and/or lack of differentiation of the cell; or it could be that a point mutation changes the regulation, the inhibition, of a c-onc so that it goes wild and produces too much of what should be a good thing: growth promoters.

Sometimes chromosomes and their sequences can become rearranged so that one may get many copies of a gene sequence, instead of only one. One can then get much more of the relevant mRNA produced than is good for the cell's behaviour (Stephen C, Fig. 59, has his abnormality because a small piece of chromosome 21 and its genetic activities became reduplicated in every body cell.) There could be multiple transcripts of a c-onc as a cause of cancer.

It could be that deletion of a segment of chromosome carrying a regulator gene might allow a c-onc to run amok. It could be—can be—that a chromosome translocation might remove from the immediate vicinity of a c-onc a regulator sequence; or a translocation might bring into the immediate vicinity of a c-onc an activator sequence, stimulating the c-onc into unwanted stimulating activities for cell proliferation.

There are about 50 known c-oncs scattered through the human chromosome complement. Every chromosome, I think, except Y have been implicated. They have, like the restriction endonuclease enzymes, strange names. Important ones are c-myc, c-abl (Abelson) which cause leukaemia and lymphomas and the 'ras' family of c-oncs that can cause bladder cancers, some forms of lung cancer and intestinal carcinomas. Let us take one or two examples of how visible chromosome anomalies can be associated with malignant disease.

For 20 years or so it has been known that chronic myeloid leukaemia and the abnormal cells involved show, replacing one normal chromosome 22, a small chromosome, the Philadelphia chromosome (Fig. 106). It is now known that there has been a reciprocal chromosome translocation whereby a fragment of the long arm chromosome 9 has been 'swapped' with a part of the long arm of chromosome 22, the piece lost from chromosome 22 being slightly bigger than the piece gained from chromosome 9; hence the small size of the reconstituted chromosome 22. The effect of this is to bring the c-onc, c-abl, from chromosome 9 into conjunction with the gene coding from the lambda(λ)-light-chain immunoglobulin on chromosome 22. In some way this association 'fires up' c-abl in such a way that cells thus affected become cells of chronic myeloid leukaemia.

Burkitt's lymphoma is another example. In the cells affected by this malignancy there has, most usually, been a translocation between the short arm of chromosome 8 and the long arm of chromosome 14; sometimes in Burkitt's lymphoma the transfer is to the long arm of chromosome 22 or to the short arm of chromosome 2. Whatever the translocation, the involvement of the long arm of 8 is essential; whatever the translocation, that translocation brings the c-onc, c-myc into conjunction with a gene coding for synthesis of a heavy-chain immunoglobulin. In some way this association 'fires up' c-myc, causing affected cells to form a Burkitt's lymphoma.

Trisomy of chromosome 12 results in chronic lymphatic leukaemia. In

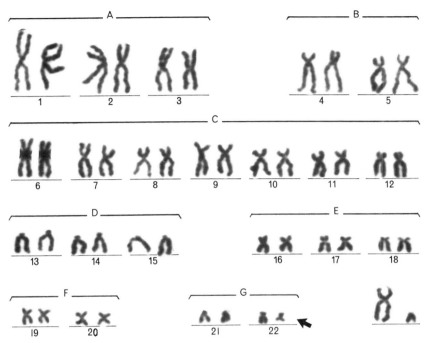

Fig. 106. *The Philadelphia chromosome, usually formed by a reciprocal translocation between the long arms of chromosomes 9 and 22; block-stained karyotype.*

acute, non-lymphatic leukaemia the chromosomal changes may be many and varied with some indication that the nature of the chromosome change may have some bearing on the prognosis. An inversion within chromosome 16 has a better prognosis than if the leukaemia is associated with trisomy 8 in the malignant cells which in turn carries a better prognosis for length of survival than if there are multiple and complex chromosome rearrangements.

In acute lymphatic leukaemia the chromosome changes are various; indeed the chromosomes of the abnormal cells may appear to be (but may not truly be) normal. If, as may happen in a small number of cases of acute lymphoblastic leukaemia, a Philadelphia chromosome is present the prognosis is very poor. Translocations involving chromosomes 8 and 4 and chromosomes 4 and 11 likewise have a very poor outlook for long-term remission. Rather oddly, if cells with extra chromosomes, 50 or more, are seen there is a very hopeful prognosis; better even than if the cell karyotype appears (but may not be) normal.

Trisomy 21, Down's syndrome, presents as yet unsolved problems in relation to the predisposition to leukaemia. It is, of course, tempting to suppose that the trisomy 21 introduces a triple dose of c-oncs as does the trisomy 12 seen in cells from chronic lymphatic leukaemia. But the situation with Down's syndrome is not comparable. The trisomy 21 in Down's syndrome is in all cells, right from the beginning. The chromosomal changes in the cancers mentioned above—and indeed in all cancers where chromosomal changes are seen—are in the cancer cells only, and not in all cells of the body. Cancers represent somatic, localised, mutational events. Something goes amiss, be it a point mutation, a specific deletion or translocation, in one cell and its proliferating descendants. The changes in cancers are somatic mutations with subsequent monoclonal growth. We do not yet know why trisomy 21 predisposes to leukaemia. It may be that the immunological surveillance, which eradicates the somatic mutational events from which we all suffer all the time, fails in the patient with trisomy 21.

Retinoblastoma, an eye malignancy of early childhood, appears to come in two varieties—one without visible chromosome abnormality and one with an (almost always) visible deletion of chromosome 13 long arm at band q14. The evidence seems to be that in the former variety all cells of the body are hereditarily heterozygous for the retinoblastoma gene but that retinoblastoma only develops when a mutational event occurs, be it a point mutation converting the normal homologous allele to a retinoblastoma gene or a deletion of the normal allele. Either way the cell has only one kind of relevant gene, the retinoblastoma gene. The frequency of occurrence of somatic mutation leading to this homozygosity must be high, for usually in this variety (that appears to be dominantly inherited, but in fact requires cellular genetic uniformity)—the tumour affects both eyes as two separate somatic mutations.

Patients with retinoblastoma with chromosome 13q14 deletion have that deletion, and presumably deletion of a normal allele at that locus, either as a *de novo*, non-hereditary event or by inheritance of the deleted chromosome. Presumably a point mutation of a specific c-onc, or point mutation of an associated sequence to promoter or enhancing activity, then causes the c-onc, unopposed because of the chromosome deletion which has removed a restraining homologue, to go wild and cause a retinoblastoma as the monoclonal descendants of the one cell in which the c-onc activation took place. A somewhat comparable situation exists in relation to the syndrome of aniridia and Wilms' tumour. In about 2 per cent of cases of Wilms' tumour, congenital aniridia is also present, and with this combination, but not with Wilms' tumour alone, there is an interstitial deletion in the short arm of chromosome 11, close to the centromere at 11p12. Any case of aniridia should have chromosome studies; if the 11p deletion is found, that child is extremely likely to have, or develop, a Wilms' tumour which is likely to be bilateral. One presumes that mechanisms similar to those cited above for retinoblastoma may account for this association between 11p deletion and the aniridia–Wilms' tumour syndrome.

With all the foregoing one can now see how one might expect to see cancers in the 'chromosome break disorders' with their propensities to chromosome and genetic rearrangements. Perhaps one can indeed say with truth 'cancer is essentially a genetic disease'.

Recommended Reading

Bishop, J. M. (1982). 'Oncogenes'. *Scientific American* **246**, 80–92.

Croce D. M. and Klein G. (1984). 'Chromosome translocations and human cancer'. *Scientific American*, **252**, 54–60.

Sparkes R. S. (1984). 'Cytogenetics of leukaemia'. *New England Journal of Medicine* **27 September**, 848–849.

Weiss R. A. and Marshall C. J. (1984). 'DNA in medicine. Oncogenes'. *Lancet*, **2**, 1138–1142.

CHAPTER 10

Sex Chromosomes, Reprise

Since in our discussions we will be freely using the terms 'male' and 'female' we must consider what those terms mean. It is less easy than it might seem. Do the terms refer to the sex chromosome complement, the nature of the gonads, the appearance of the visible genitalia or to how a person is perceived to function in society?

When I use those terms I refer to the phenotypic sex, what the genitalia look like—or most look like. The phenotypic sex characteristics determine that instant Oslerian diagnosis, made within seconds of birth: 'It's a boy!' Phenotypic sex is diagnosed irrespective of what gonads there may be or what chromosome complement exists. A male, then, is one who looks, to the observer of the external body form, like a boy or man (Fig. 107). He has at least a penis. A female looks like a girl or woman. There is a urogenital cleft. The ability to produce sperm or ova, to menstruate or to be fertile are irrelevant to our definition here.

Because of the rather odd economy of nature that uses the same apparatus for both reproduction and urination (an economy that might be explained by the supposition that passages that are relatively infrequently used might benefit from frequent flushings) sex, in our terms of reference is highly relevant to what toilet facilities are used. We might almost say that a female is one who sits with the ladies; a male one who stands with the gentlemen. It certainly is usually so.

The point is not made frivolously, for, from a practical viewpoint an individual must do one or the other in our society. The anatomical endowment will determine the appropriate choice—at least, it should. Indeed, irrespective of dress or hairstyle, the call of nature will be the moment of truth (Fig. 108). If, in our society, a man must of anatomical necessity sit like a lady he would suffer agonies of embarrassment and, most likely, cruel ridicule. It would be most unrealistic and most unkind to raise a child as a male simply because his chromosome complement is XY and his gonads histological testes if in the school locker room he looks like a girl and is unable to use a male urinal.

Whether psychic sexual orientation is entirely undifferentiated at birth is still in some doubt, but increasingly the evidence is that prenatal

There <u>IS</u> a difference !

Fig. 107. There is a difference in phenotypic sex (adapted by Leonie Duncan from a beer mat in the possession of Professor Murray Barr).

hormonal influences may, to some degree, 'programme' the brain, throughout life, to 'think male' or 'act female', though upbringing and the sex of rearing are much more potent influences. It has been thought that gender role and the light in which the child sees himself/herself is established irrevocably by 2 years of age or thereabouts, but more recent experience with sex-reversal procedures and physiological reversals make it seem that acceptance of a gender role is less irrevocable than has been thought. However, the greatest caution must be exercised before risking the psychic turmoil that may result from a late change of sexual identity. Perhaps the recognition of reproductive potential in a role different from that assigned might justify a change but a well-adjusted gender role discordant with genetic sex is better than mental anguish caused by demands of scientific tidiness and sexual congruence.

Sex Determination

Although the factors determining sex differentiation in the embryo have

There _IS_ a difference!

Fig. 108. There is a difference in psychosocial adaptation (drawn by M. L. Jones, Department of Instructional Resources, University Hospital, London, Ontario, Canada).

been briefly discussed early in this book, a restatement may be helpful here.

The genetic sex is the resultant of more than one gene, or so it seems. A gene on the X chromosome, together with one (the only one) on the Y chromosome and together with, perhaps, a gene on an autosomal chromosome, directs the production on cells of the germinal ridge of an inducer substance which seems to be identical with a protein known as the HY antigen. Under the influence of this HY antigen the cells of the germinal ridge differentiate to form testes. If no HY antigen is made, either because there is no Y chromosome or because the other relevant genes are absent, the cells differentiate as ovaries.

At first the developing testis is stimulated to grow under the influence of placental chorionic gonadotropin. Later, as the pituitary assumes its functions the control passes to pituitary gonadotropin.

The testis has two kinds of hormone-producing cells: Sertoli and Leydig cells. The former produce a hormone, the Mullerian duct regression factor, that causes the organs developed from that duct to cease development and to wither away. The Leydig cells, by several metabolic steps, produce testosterone. This testosterone stimulates development of the Wolffian duct so that epididymis, vas deferens and seminal vesicles are formed. Continuing activity of testosterone causes growth and maturation of the external genitalia as puberty approaches and growth of muscles, deep voice and libido. It may influence, prenatally and later, thought processes.

Some of the testosterone is changed by the enzyme, 5-alpha-reductase to dihydrotestosterone which is taken up by receptors of the cells destined to form the external genitalia. Within the cell the dihydrotestosterone so regulates the cell proliferation and differentiation that the male external genitalia are appropriately formed. There is some reason to believe that continuing dihydrotestosterone activity may cause acne, baldness and prostatic hypertrophy (Fig. 109).

If the zygote receives no Y chromosome, or if the other genes relevant to the production of HY antigen are lacking, the germinal ridge develops as ovaries. No Mullerian duct regression factor is made, the uterus and tubes develop, testosterone is not made, nor is dihydrotestosterone. There is no development of the Wolffian ducts to epididymis, vas and vesicles and no masculinisation of the external genitalia. A female is, if you like (but my wife would not!) a female—by default; but not quite. Mere absence of a Y chromosome does not ensure development of normal ovaries. Two X chromosomes or more are essential for normal ovarian development: a 45,X chromosome complement allows only 'streak' ovaries. A 46,XY chromosome complement, with failure of the supporting genes to produce HY antigen, does not allow normal ovarian development. There is, as with 45,X, ovarian dysgenesis.

In general (and there are always exceptions) homogeneous sex chromosome complements do not cause ambiguous genitalia, though mosaicisms may do so. The exceptions may be caused by unrecognised mosaicisms or mosaicisms from which in embryonic development one stem-line has died out leaving homogeneity of the complement.

Masculinisation of an XX Fetus

It is not only by testosterone produced by the fetal testis that differentiation as a phenotypic male can come about. If the mother were to take

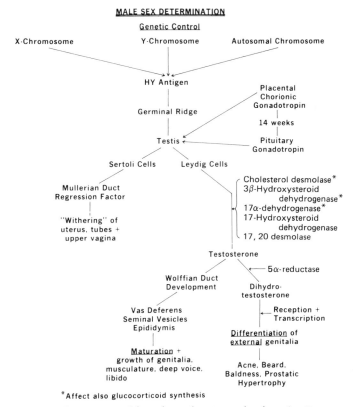

MALE SEX DETERMINATION

Genetic Control

X-Chromosome   Y-Chromosome   Autosomal Chromosome

HY Antigen

Placental
Chorionic
Gonadotropin

Germinal Ridge

14 weeks

Testis ←   Pituitary
                            Gonadotropin

Sertoli Cells   Leydig Cells

Cholesterol desmolase*
3β-Hydroxysteroid
                dehydrogenase*
17α-dehydrogenase*
17-Hydroxysteroid
                dehydrogenase
17, 20 desmolase

Mullerian Duct
Regression Factor

"Withering" of
uterus, tubes +
upper vagina

Testosterone

5α-reductase

Wolffian Duct
Development   Dihydro-
                     testosterone

Vas Deferens
Seminal Vesicles   Reception +
Epididymis   Transcription

Differentiation of
external genitalia

Maturation +
growth of genitalia,   Acne, Beard,
musculature, deep voice,   Baldness, Prostatic
libido   Hypertrophy

*Affect also glucocorticoid synthesis

Fig. 109. *Determinants of the male sex phenotype and male sex function.*

testosterone or some other androgenic hormone in pregnancy there would be conversion of that androgen to dihydrotestosterone and there would be masculinisation of the external genitalia; but there would be no inhibition of ovarian development, and no regression of Mullerian ducts. Ovaries, uterus and fallopian tubes would continue to develop.

The fetus might produce androgens intrinsically, in error, for itself. If in the metabolic pathway for the synthesis of cortisol, hydrocortisone, there should be a block to further synthesis, there would be accumulation of metabolites, 'backed-up', proximal to the block. In Fig. 110 one illustrates two such blocks, two such inborn errors of metabolism: blocks at 21-hydroxylation and at 11-hydroxylation. In this situation 17-OH-progesterone and 17-OH-pregnenolone are the backers-up. Their conversion to androgenic, testosterone-like, androsterone can lead to exter-

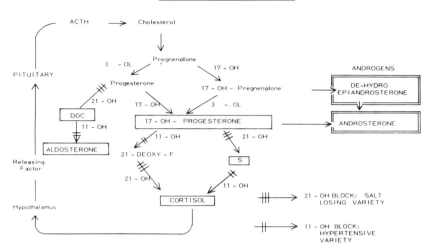

Fig. 110. Biochemical pathways in the synthesis of cortisol, metabolic blocks and the alternative routes whereby accumulated intermediate products may be converted to androgens.

nal masculinisation, to greater or less degree, of the external genitalia (Fig. 111).

While such is not a chromosome disorder, it is mentioned here because there is a disparity between the phenotype and the chromosome complement. Moreover, it is one of the few disorders in which a chromosome study can be a matter or urgency. In Fig. 110 one sees to the left-hand side of the diagram that a 21-OH block causes failure of synthesis of the mineral retaining steroids, deoxycorticosterone (DOC) and aldosterone. In that situation there can be an acute crisis of life-threatening Addison's disease with hypotension, hyponatraemia and dangerous hyperkalaemia. In a baby with ambiguous genitalia as in Fig. 111 and a shock-like state it can be most important to know that the baby has an XX constitution with masculinisation: that it has salt-losing virilising adrenal hyperplasia. This is one of the few occasions when a buccal mucosal smear is used to give an answer in minutes; is there a Barr body present (XX) or not (XY or 45,X)?

The X Chromosome and Inactivation

In Chapter 2 we discussed the important property of X inactivation whereby at about 2 weeks after conception, randomly, in each and every

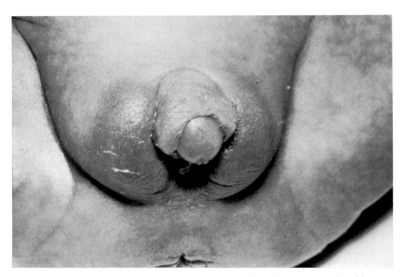

Fig. 111. Susan C: masculinisation of a female fetus, XX, with virilising adrenal hyperplasia of the salt-losing variety.

cell developed up to that time, one X chromosome becomes partly inactive and in the state is visible as the Barr body. We mentioned the n-1 rule, recapitulated in Fig. 112. The inactivation of any X in excess of one cannot be complete. If it were, an XX female would be no different from one who has only one X, a 45,X complement; but she is.

If a woman is heterozygous for a gene on the X chromosome, that is to say if she has on one chromosome one allele and on the other one which gives different instructions, she will be a genetic mixture: a mosaic. The random inactivation of one or other X, lyonisation, will lead to, genotypically, two cell populations. In one population one allele will be giving genetic instruction, in the other, the other allele. The effect of that mosaicism can be seen in the woman who is, let us say, heterozygous for the gene for glucose-6-phosphate-dehydrogenase (G6PD) deficiency (which, among other effects, can cause an haemolytic anaemia after eating broad (fava) beans, favism). She can be shown to have two populations of red cells—normal and those showing G6PD deficiency—but she will not show any clinical ill-effects.

Suppose she has a son who has been begotten of an ovum which at meiosis of gametogenesis received the X chromosome with the allele coding for G6PD deficiency, that son will have only one X, the 'bad' one in all his cells. He will be hemizygous for the 'bad' allele. He might

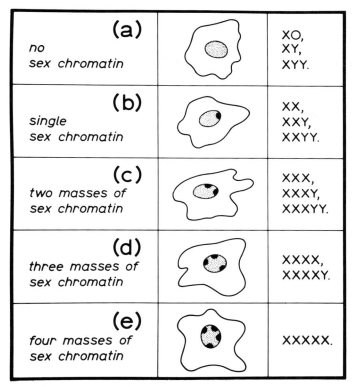

(a) no sex chromatin		XO, XY, XYY.
(b) single sex chromatin		XX, XXY, XXYY.
(c) two masses of sex chromatin		XXX, XXXY, XXXYY.
(d) three masses of sex chromatin		XXXX, XXXXY.
(e) four masses of sex chromatin		XXXXX.

Fig. 112. The n-1 rule; the number of Barr bodies in relation to the number of X chromosomes.

become quite sick if he were to eat Windsor broad beans, take antimalarial drugs or even take aspirin.

Or suppose a woman has, by error of chromosome segregation at gametogenesis in one or other of her parents (usually the father), only one X chromosome and that were to carry the 'bad' allele; apart from having ovarian dysgenesis, she too would be hemizygous for the 'bad' allele and could show the effects of an X-located recessive gene—just like a man. In this way a phenotypic female with a 45,X (sometimes called XO) constitution can have haemophilia, Duchenne muscular dystrophy or favism.

The Y Chromosome and X Inactivation

As we have said above, this chromosome seems to have one genetic preoccupation—sex. Whether the Y chromosome carries the structural gene that codes for the HY antigen, or whether it carries a regulator gene that acting on a structural gene on an autosome is really an irrelevant detail, though it is quite probably the latter. The presence or absence of the Y chromosome is the arbiter of sex determination. The locus of the gene determining fetal testicular development is located on the short arm, Yp. The function of the long arm, if any, is unknown. It is so brightly fluorescent, with quinacrine staining, and remains so tightly compacted, even in an interphase cell, that it can be seen as a brightly glowing dot in buccal mucosal or blood cells (see Fig. 18). Each and every Y chromosome can be seen. If there are two, two Y bodies will be seen (see Fig. 19).

Sex Chromosome Mosaics

Mosaics of the sex chromosomes are much more common than, let us say, mosaics of trisomy 21: Down's syndrome mosaics. This is probably because sex chromosome anomalies, even 45,X are not very lethal to the cells that carry them. If there were to be a mosaicism thus, 45,X/47,XY, + 21 or 46,XX/46,XY or 46,XY/47,XXY, all these stem-lines are likely to survive and be found on cytogenetic investigation. Triple, quadruple and multiple stem-line mosaicisms of sex chromosomes are on record. Such is not the case with autosomal chromosomes.

CHAPTER 11

Chromosome Anomalies in the Female

Ovarian Dysgenesis

Although Morgagni first described congenital ovarian deficiency in 1761 it is only recently that there has been much understanding of the condition.

It is better to use the term dysgenesis rather than ovarian aplasia for it is rare that no ovarian tissue is found. Usually it is the germ cells that are especially lacking, but even this lack may be more relative than absolute. In many cases germ cells may be present in the fetus with diminution in their number as time goes on, so that by adult life few, if any, remain.

We are all familiar with failure or error of organ development for reasons which, while often unknown, are not believed to have come about from genetic or chromosomal cause. We see cases of renal dysgenesis, microphthalmia, dysgenesis of the lung. It is not difficult to imagine that some cases of ovarian dysgenesis might have arisen from vascular accident, viral infection or some other noxious circumstance in embryonic life. One can imagine ovarian dysgenesis as an isolated malformation without genetic or chromosomal defect: 'pure gonadal dysgenesis'.

We have already said that, although one X chromosome is partly inactivated, two are necessary for entirely normal development. This is especially true of ovarian development. Loss of a whole, or even a part of an X chromosome leads to ovarian dysgenesis of greater or less degree.

But chromosome anomaly is not the only cause of ovarian dysgenesis related to genetic malinstruction. Suppose one has an XY zygote and early embryo, but suppose also there were to be a single-gene defect either of the Y chromosome or a defect of other genes which, in conjunction with the Y chromosome determine testicular development, what would happen? The germinal ridge would not develop as a testis for there would be no HY antigen to make it do so. It would be more destined, then, to develop as an ovary. But we have said that two X chromosomes are required for normal ovarian development. But we have also said that this zygote and embryo is XY: the Y can do nothing; the one X is insufficient for normal ovarian growth and development. We get ovarian dysgenesis in an XY individual. The uterus and fallopian tubes may develop, though often imperfectly. The external genitalia will, of course, remain female and unmasculinised. So we will have a female, with gonadal dysgenesis but an XY chromosome constitution: 'mixed gonadal dysgenesis'.

Turner's Syndrome, 45,X or XO

Undoubtedly, rare though it really is, the best known variety of gonadal dysgenesis is that associated with Turner's syndrome: failure of germ cell maturation (or early degeneration) combined with short stature, a number of minor and inconstant defects and a chromosome complement lacking either a whole X chromosome or lacking a major amount of the short arm of one X chromosome. One has either a 45,X chromosome complement, or by simple deletion or isochromosome formation, loss of the short arm of X, which seems to be the most important so far as the typical features of the syndrome are concerned; one can have, therefore, Turner's syndrome as 45,X or as 46,XXp − or as 46,X,i(Xq).

Loss of the long arm of an X chromosome certainly gives ovarian dysgenesis (for two complete and entire X chromosomes are required for normal ovaries) but we do not get short stature nor the other characteristic features. Factors for normal stature seem to reside in the short arm.

While Turner's syndrome is not ordinarily a familial disease, there is more than one report of two affected sibs. We do not know if there may be in some parents an especial liability to malsegregation of a sex chromosome. We do know from the pioneer work of Carr and subsequent studies by others that living cases of Turner's syndrome with a 45,X chromosome complement are but the small residue of those that are conceived. Perhaps 98 per cent of 45,X fetuses are rejected as miscarriages; only 2 per cent survive as fetuses to be born alive. Whereas 5 per cent of aborted fetuses are 45,X, no more than about one in 7000 living females has Turner's syndrome. It is very strange indeed that a 45,X complement is so lethal to the fetus but not so beyond the first half of pregnancy.

The condition is not easy to diagnose in the newborn, but there may be some features that might arouse suspicion. There may be peripheral oedema, lymphoedema, especially of the feet (Fig. 113) or, even if there is not webbing of the neck, the skin of the neck may be loose and redundant (Fig. 114); even in infancy it might be noted that the nipples are widely separated and the chest broad with a short sternum. The late David Smith of Seattle, the dean of dysmorphologists, told me that he

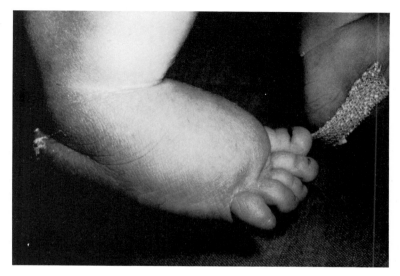

Fig. 113. Lymphoedema in a newborn with 45,X Turner's syndrome.

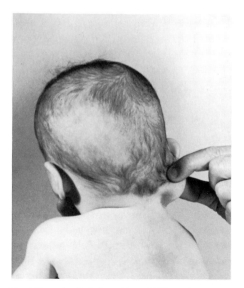

Fig. 114. Loose skin of the neck of a baby with 45,X Turner's syndrome; if this skin had been more redundant there might have been webbing of the neck.

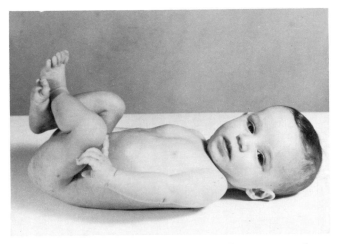

Fig. 115. Laura B was diagnosed as having a 45,X chromosome complement only because she was found to have coarctation of the aorta.

had come to regard sparse temporal hair as a suggestive feature even in the newborn nursery. Laura B (Fig. 115) does perhaps show this feature.

Congenital cardiovascular defects involving the left side of the heart are rather common and coarctation of the aorta—quite rare in females without Turner's syndrome—is found in about one-quarter of cases. Where the diagnosis is suspected the femoral pulses should be sought. If they are at all weak the blood pressure in the arms and legs should be measured. Baby Laura was referred at 7 days of age for cardiac failure. The femoral pulses could not be felt. The blood pressure in the arms was extremely high at 160 mmHg systolic pressure. In the legs it was 60 mmHg. The baby was found to have a 45,X chromosome complement. The coarctation has been resected. Had it not been for the cardiac failure and the coarctation the diagnosis would have been missed and the baby passed without qualification as suitable for adoption, as was the intention.

Even in later life the diagnosis is not always obvious. The lymphoedema disappears, the broad chest can easily pass unnoticed and the loose folds of skin of the neck, quite often absent anyway, may also pass unnoticed unless a distinctive web is formed (Fig. 116). However, shortness of stature is universal. Any female who for no obvious reason of sickness or heredity is below the third percentile of normal height for her age (or is two standard deviations below mean height) should be con-

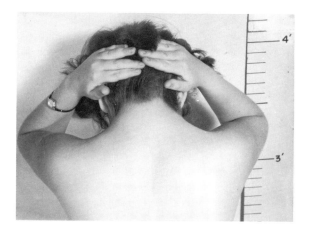

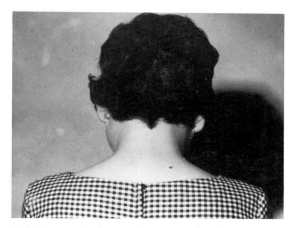

Fig. 116. Girl A has a webbed neck; girl B does not, but the contour is slightly flat. A deeply pigmented naevus can be seen.

sidered as a possible case of Turner's syndrome and other specific features should be sought.

Even if there is no webbed neck, the hairline may be low and the neck broad, and there may be an unusual number of dark pigmented naevi (Fig. 116). There may be cubitus valgus, an increased 'carrying angle' at the elbow as in the girl on the left of Fig. 117. It is not a very distinctive sign being commonly present also in normal people. If the fist is clenched it may be seen that the 4th metacarpal is short. The dermato-

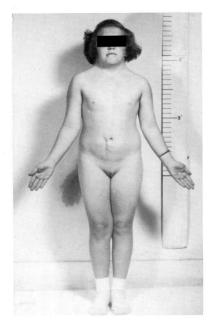
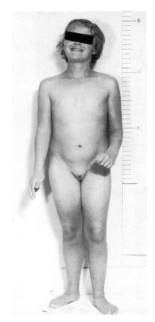

Fig. 117. The girl on the left shows the rather typical body build of Turner's syndrome; the increased carrying-angle is well demonstrated. The girl on the right also has a 45,X chromosome complement but with normal build; she has no pubertal development at 17 years of age.

glyphics are not strikingly abnormal though the axis triradius may be distally located. The finger nails are often narrow and hyperconvex. The skin tends to form ugly keloid scars which should be borne in mind if cosmetic surgery is contemplated. Congenital heart disease, involving especially the aorta, should be looked for; coarctation of the aorta is common in Turner's syndrome, but is rare in normal girls.

Ultrasonography or a pyelogram will often reveal renal abnormalities; classically there is a horseshoe kidney. Some patients have a Meckel's diverticulum, a rare finding in normal girls. This might cause massive painless melaena.

One may find few—or even none—of the above features, but, even so, a very short girl deserves a chromosome study even if there are no stigmata of Turner's syndrome.

Very often it is only when the expected pubertal changes fail to appear that there is real suspicion that something is amiss. The breasts fail to develop, pubic and axillary hair is very scanty or absent and there is no menstruation (Fig. 117). The external genitalia remain infantile and the vaginal smear shows no oestrogen effect. Laparoscopy shows the characteristically small, fibrous 'streak' ovaries. The uterus and tubes are small but normally formed. Sterility, of course, is a virtual certainty in the case with a homogeneous 45,X chromosome complement, but there may be fertility in some cases of mosaicism or cases in which there is loss of part only of an X chromosome. Since the defect is of ovarian dysgenesis it seems that with appropriate hormonal replacement and an ovum donation pregnancy might be achieved. It is not impossible.

That all this, and the short stature, is not due to pituitary failure is certain. Indeed pituitary gonadotropins are at unusually high levels; it is as though the pituitary, taking cognisance of low ovarian activity, is trying to whip into action a flagging gonad. As a result of this increased gonadotropin activity, 'hot flushes' may be experienced.

What of the intellect and personality of the Turner's syndrome patient? It has been commonly believed that these patients are quiet, withdrawn and often retarded. This is not true. If anything, they tend to be bright, lively, vivacious, often a leader of their peer group. Few are retarded. One study has shown that as many girls with Turner's syndrome go on to university as their normal sibs.

It is also said that these girls have specific difficulties with spatial relations, drawing and directional descriptions, that they easily become lost in a large building or city streets. From some observations, I doubt whether this is true.

Cytogenetic Features of Turner's Syndrome

In the majority of patients with classic Turner's syndrome the chromosome complement is 45,X or as it is sometimes written, XO (Fig. 118): monosomy of X. With the possible exception of very rare cases of monosomy 21, this is the only homogeneous monosomy compatible with life. Even 45,Y or YO is invariably lethal to the very early embryo.

There is evidence that in the majority of 45,X patients the monosomy arises from lack of the paternal contribution. In about 80 per cent of cases the solitary X is of maternal origin. The incidence of Turner's syndrome is not related to either maternal or paternal age.

The next most common cause of Turner's syndrome is loss of the short arm of X by the mechanism of isochromosome formation, resulting in a 'double dose' of the long arm of one X chromosome, trisomy of the long arm, and absence of the short arm: monosomy of the short arm of X (see Fig. 43). This is 46,X,i(Xq).

Some cases are due to simple deletion of the short arm of X:

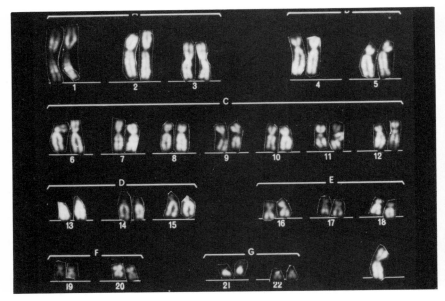

Fig. 118. Karyotype of 45,X complement.

46,X,Xp − or 46,XX,del(X)p. Some cases, as we have said, may be due to mosaicism of varying degrees, with the usual complement being 45,X/ 46,XX. The complement 46,XXq − or 46,XX,del(X)(q) usually is asso- ciated with normal stature and few, if any, somatic stigmata, but there is gonadal dysgenesis. Two entire X chromosomes are required for normal ovarian development.

The reader will see that, with some cases being due to isochromosome formation, others to partial deletion of the X chromosome and others to mosaicism, the use of a buccal mucosal smear for Barr body examination could be misleading. Barr bodies could be found, and yet the patient could have Turner's syndrome. Where there are clinical grounds for sus- picion of Turner's syndrome, a cell culture and karyotype analysis is required.

Clinical Management of Turner's Syndrome and Gonadal Dysgenesis

Matters involving sexuality and sexual function require most delicate management. These matters are charged with great emotional content. It would be a cruel error to tell a girl, because she has only half the normal number of sex chromosomes, that she is only 'half a woman' or that she is 'halfway' to being a man. She is a girl, or a woman, in whom the ovaries have not developed. The reason why they have not developed is really of no great importance; unless demanded, cytogenetic explanations may be best avoided. It is perhaps better to point out that there are people born without thyroid glands, with a kidney missing, and that there are people whose pancreas does not function properly, who require hormone replacement therapy.

One can then explain that the failure of pubertal development is due to the failure to produce the requisite hormones but that these can be given as pills, a great deal more easily than insulin to replace pancreatic deficiency. Oestrogens are the mainstay of therapy, but cyclical therapy with progestins may be used to produce menstruation. This greater normality may have psychological advantages—or might be regarded as a 'curse'. Small doses of androgenic hormones may be added. One may, but also may not, gain a little extra height that way. One cannot offer normal stature.

One need not rush to tell a young girl that she will not be able to have babies of her own—unless she asks; and nowadays some children will. One might start inducing pubertal changes at, let us say, 12 years of age. The news about reproduction could wait for a year or two. The time will come, however, when the truth must be told, certainly before serious dating and the possibility of engagement or marriage is in the immediate future. It must now be explained that, not only are the ovaries not making hormones but that it does not make egg-cells either. This cannot be put right. One cannot offer fertility. One must make sure that she understands that her disease, if treated, is no barrier to full sexual satisfaction for both parties and that she will not be frigid and unresponsive. It is, of course, only fair that an intended husband should know that she will be unable to have children of her own. Such a phrase is less threatening than 'sterility', 'infertility' or 'barrenness'. Needless to say one will point out that very many couples are in the same position and that adoption will give them joy and happiness. Whether one would point out that a careful recent study showed greater matrimonial happiness among childless couples than among those with children might be a matter of opinion!

In this type of gonadal dysgenesis there is no real danger of malignant change in the 'streak' ovaries, but it is otherwise in the other varieties of gonadal dysgenesis. Those cases, like cases of Turner's syndrome, will require hormone replacement therapy but cases of gonadal dysgenesis with a Y chromosome (even though ineffective) in the complement should have the dysgenetic gonads removed. There is a high risk of a malignant gonadoblastoma.

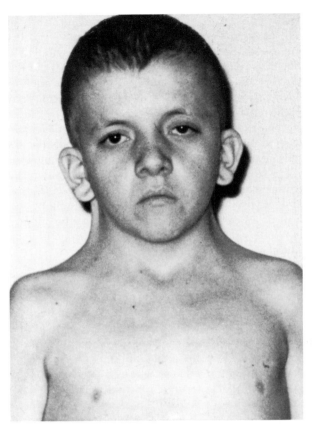

Fig. 119. Noonan's syndrome, sometimes confusingly and incorrectly called 'male Turner's syndrome'.

'Turner's Syndrome' in Males, Noonan's Syndrome

Occasionally males are seen with short stature, webbed necks, congenital heart disease and other stigmata that are features of the 45,X chromosome complement in females (Fig. 119). With rare exceptions (for example, 45,X/46,XY mosaicism) these males have a normal chromosome complement, 46,XY. They are 'phenocopies' or imitators of Turner's syndrome, cases in which the same result has been arrived at by a different mechanism. These phenocopies are cases of Noonan's syndrome. One has to admit that within that classification there is heterogeneity. Some cases of Noonan's syndrome are inherited as autosomal dominants with variable degrees of expression in different individuals; some appear to be due to recessive inheritance. In some the mechanism is not clear. Females can have Noonan's syndrome. They will resemble cases of Turner's syndrome, but the chromosome complement will be 46,XX. It is said that in Noonan's syndrome the congenital cardiac defect, if one is present, is right-sided, pulmonary stenosis for example, rather than left-sided as in Turner's syndrome.

Testicular Feminisation Syndrome, Androgen Unresponsiveness

This is not a chromosome disorder, but it is discussed here because there is a disparity between the phenotypic sex and the chromosome complement. The baby and young child looks to be an entirely normal female; but the chromosome complement is 46,XY. In Fig. 109 we showed that testosterone produced by the fetal testes (developed under the influence of the Y chromosome and its assistant genes) is, in part, changed by the enzyme 5-alpha-reductase to dihydrotestosterone which is taken up by cells destined to differentiate into the external genitalia. If the dihydrotestosterone is 'received' into the cell's mechanism and its hormonal message transcribed appropriately the cells will differentiate in such a way that the genital tubercle of the, as yet, undifferentiated genitalia will grow as a penis; the urogenital sinus will close over so that a urethra is formed that, in due course, will extend to the tip of the penis. The uterus and tubes will, of course, wither away under the influence of the Mullerian duct inhibitor.

There is a very rare disorder determined by autosomal recessive genetic error in which 5-alpha-reductase is deficient; thus no dihydrotestosterone is made and the genitalia remain female even though testes are present and testosterone produced. Very few families with this disease are on record. Much more common is the testicular feminisation syndrome (TFS).

In this syndrome, an X-linked (X chromosome located) gene gives malinstruction to the receptor cells of the fetal genitalia so that they do not take up, receive and incorporate into the cell function the dihydrotestosterone: receptor failure.

A female heterozygous for this defective gene will have little or no effects, though pubic hair may be rather scanty. In her the normal homologous gene outweighs the effect of the abnormal gene. She is a symptomless carrier: just like a female carrier of the genes for haemophilia or Duchenne muscular dystrophy. Any daughter, XX, will have received

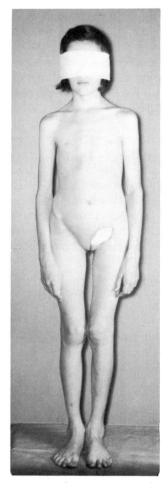
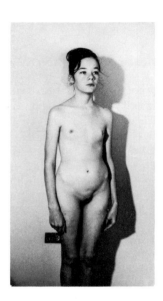

Fig. 120. Testicular feminisation syndrome. On the left at age 12 this girl was found to have bilateral inguinal lumps. At first these were thought to be hydroceles of the canal of Nuck, but they were found to be testes and removed. There is no uterus, the vagina is about 5 cm long. On the right she is aged 14 and has been taking oestrogens for 18 months. Breasts and pubic hair are developing.

one X chromosome from her father; from her mother she will have received either the 'good' or the 'bad' gene. Any daughter will be at a 50 per cent risk of being a carrier of the TFS gene.

A zygote with a 46,XY complement will be hemizygous for the TFS

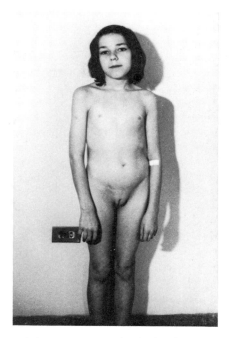

Fig. 121. Sister to girl shown in Fig. 120; she also has the testicular feminisation syndrome. An inguinal hernia was repaired in infancy and a testis found and removed. She has not yet started on hormone therapy.

gene; the gene will be unopposed. The Y chromosome and its assistants will do their parts. Testes will develop, testosterone will be produced, vas deferens and seminal vesicles will develop and the Mullerian ducts will wither away. So far, so good. But the cells of the embryonic external genitalia will not respond to the dihydrostestosterone. The genitalia will remain female, but inside there will be no uterus nor tubes. The testes are often in the inguinal canals or they may be in the abdomen. So one has a girl baby with lumps in the inguinal canals, which might be hernias, cysts of the canals of Nuck, or may well be testes. Unless one is quite sure of the nature of inguinal lumps, a chromosome study is in order, as it might show 46,XY.

What does one see in a family pedigree in this disorder? One sees a preponderance of females. The XX females will be females, of course. But, equally, half the XY individuals will be called females. One may see childless maternal 'aunts' who never had menstrual periods; these aunts (for such they would be called) would have a 46,XY chromosome consti-

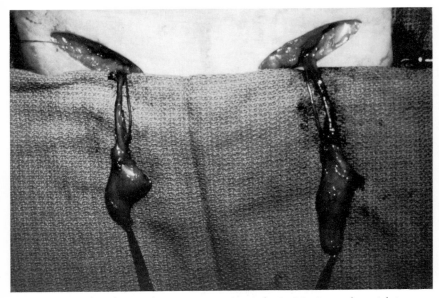

Fig. 122. Bilateral inguinal testes in a case of testicular feminisation syndrome (photograph at operation kindly supplied by Dr Don Marshall).

tution. One may of course see two 'sisters' with the same condition, for that is what they will be called; that is how they will appear to be (Figs 120, 121).

If nothing is done, if the condition is unrecognised, as it might well be (Fig. 121) for several years no harm will come. But at the time of expected puberty there will be no pubic hair, no menstruation, but there may be good, even bounteous, breast development. The pituitary produces a large output of gonadotropins to which the testes respond. There is a great production of testosterone, but also of oestrogens. These patients, tall and sumptuously endowed are, I have been told, to be found among the ranks of showgirls. We have, unfortunately, no such picture for this book!

Women with this condition are tall, good-looking, forceful and ambitious. There may be increased libido. The uterus and tubes, if any, are very underdeveloped. The vagina is somewhat short and ends blindly for there is no cervix.

While there is some debate about timing of operation this author feels strongly that as soon as the diagnosis is firmly established, the testes should be removed (Fig. 122). They must be removed at some time for

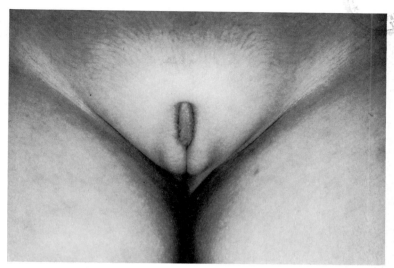

Fig. 123. Incomplete testicular feminisation syndrome due to 'sick' androgen receptors as distinct from complete receptor defect. This is one variety of male pseudohermaphroditism; this patient has a 46,XY chromosome complement.

Fig. 124. Karyotype 47,XXX.

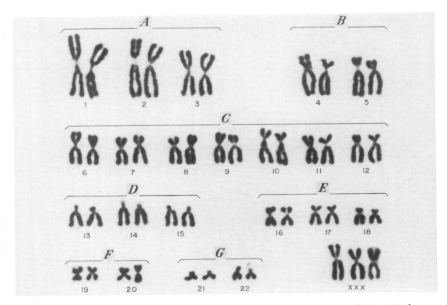

Fig. 125. Janet M, age 8 years, has a 47,XXX chromosome complement; she is a North American Indian. Her XXX complement was found when she was investigated for mild mental retardation. Not all XXX persons are retarded, and many go unrecognised.

there is a risk of malignant change, and it seems better to get such surgery over and done with rather than to invite too much discussion of the nature of the lumps and the reasons for their removal at an age of greater understanding. Little scars from surgery in infancy will not excite too many awkward questions. To tell an adolescent girl that she must now have surgery to remove her testes seems to me likely to be traumatic. The argument in favour of this late surgery is that it is best to let breast development come about on its own, rather than be induced by pills. I find that reasoning weak. In any event, after adolescent surgery, exogenous oestrogens will be required.

It goes without saying that the patient should not be told that she is anything but a girl or woman. The fact that she has, or has had, testes may be, in my view, kept secret unless there should be direct questioning. Mention of chromosomes should be avoided. Even children nowadays know that the Y chromosome is 'male'. She might be told that the womb has not developed normally, that the sex glands were abnormal and would not have functioned normally and that they were removed to guard against the risk of cancer. Provided that the vagina is adequate, as

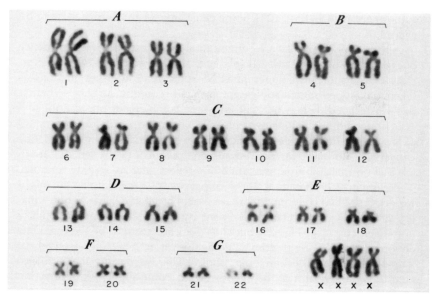

Fig. 126. Tetra X complement; 48,XXXX.

it usually is, there will be no obstacle to normal sex relations. Female sibs should have their chromosomes examined.

Sometimes the failure of reception of dihydrotestosterone is not complete. Some activity remains. There is, if you like, a spectrum of more or less 'sick' receptors with effects varying from something approaching complete failure of masculinisation (Fig. 123) to near normality.

Triple X, Tetra X and Penta X (XXX, XXXX, XXXXX)

In these disorders, by non-disjunction at the first meiotic division, at the second, or at both (see Fig. 47) two, three or four X chromosomes are present with the gamete. Fertilisation may add another X or a Y. It is believed that it is usually the female gamete that is at fault, though the possibility of an error in spermatogenesis is not impossible.

These patients are, as we would expect, chromatin positive. They have two, three or even four Barr bodies depending on whether the complement is XXX (Fig. 124), XXXX or XXXXX (Fig. 112). The 2–1 rule holds good.

Although much more frequent than Turner's syndrome, the XXX female is quite likely to pass unrecognised for often there are no phenotypic abnormalities (Fig. 125). Some cases have been diagnosed quite by chance.

Although the majority are quite normal there is an increased incidence of congenital heart disease, dysplasia of the hip and such minor abnormalities as a short little finger. Janet M, some years after her picture was taken, developed severe adolescent scoliosis requiring surgery. Whether this was related to her XXX chromosome complement one cannot say. As a young adult Janet is now a pleasant and quite pretty girl with an IQ of around 75. Pubertal development has been normal and she would, one expects, be normally fertile. Perhaps one quarter of XXX females are in the dull normal or borderline range of intellect; few are frankly retarded. There is a greater than general risk of psychiatric breakdown. One might expect, if she had children, that there would be an even chance of normality or of XXX daughters or XXY sons by the mechanism of secondary non-disjunction (see Fig. 46), but such does not seem to be the case. The abnormal gametes are discriminated against or are rejected selectively at gametogenesis and become polar bodies. There is, despite the harmlessness to life and health of the liveborn child, some slight risk of rejection of the XXX fetus. Occasionally a triple X fetus is found as a spontaneous abortion.

Tetra X (XXXX) (Fig. 126) patients are rare but are invariably retarded. Penta X (XXXXX) is, to say the least, extremely rare. These patients are small, with rather vague facial dysmorphic features. They are likely to be at least as retarded as the XXXX girls.

CHAPTER 12

Sex Chromosome Anomalies in Males

Klinefelter's Syndrome, XXY

In 1942 Klinefelter and his associates described a syndrome of aspermatogenesis and gynaecomastia in males. The name Klinefelter's syndrome is in general use, though 'testicular dysgenesis' and 'chromatin-positive micro-orchidism' are suggested alternatives. In 1956 Bradbury and his co-workers, and Plunkett and Barr (London, Canada) showed that these patients, male though they undoubtedly were, were chromatin-positive on buccal mucosa smear. In 1959 Jacobs and Strong showed that the chromosome complement was indeed XXY.

The incidence of this disorder is about one in every 750 males. In a study of boys in a school for the mentally handicapped a much higher prevalence was found, nearly ten times that figure.

The disorder is not likely to be recognised in the newborn or young child. To be sure the penis may be somewhat small and there may be mild hypospadias; the testes too may be rather small. These are rather subtle signs. There is much normal variation.

Since there is evidence that some adverse features may be modified by drug treatment it is becoming important to try and make a clinical diagnosis or at least entertain enough suspicion to order a chromosome study. A small penis, hypospadias and small testes deserve investigation.

As the child grows the body build may become unusual, but often not. There may be tall stature with a thin asthenic build or there may be a heavy build and gynaecomastia (Fig. 127) and a slight risk of breast cancer. There may be no really unusual features.

The intellect is usually normal (in contrast to earlier beliefs that intellectual impairment is common). In cases with intellectual impairment the IQ is borderline only, not profound.

It is the personality of these patients that causes their greatest problems but is, according to Smith, amenable to treatment. These patients are shy, apprehensive, passive, lonely, self-effacing but given to petty crimes, perhaps as a compensation for their feelings of inadequacy and paucity of friends. They seem, untreated, to be exceptionally vulnerable

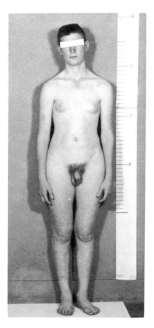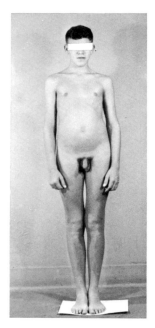

Fig. 127. Klinefelter's syndrome: note the different body builds and the gynaecomastia of the left-hand youth; secondary sex characteristics are often quite well developed though the testes will be small.

to psychic breakdown and may end up in institutional care. Although sexual drive and potency tend to be reduced there seems to be an increased risk of sexual deviations.

In adolescence and adulthood the penis is commonly within normal limits for size or only slightly small. The pubic hair is sparse and of female distribution. The beard is slow to develop and is sparse. The voice may remain feminine in pitch. There is, in brief, evidence of diminished testosterone activity and, indeed, on estimation, diminished circulating testosterone levels.

While, prior to puberty, the testes are scarcely noticeably abnormal, beyond that time they are clearly unusually small. They may be either very soft or hard.

The changes in the testes seem to increase with puberty under the influence of the pituitary gonadotropins which are always at a high level. There are no very striking histological changes before that time. The seminiferous tubules become narrow or obliterated and hyaline sclerosis is characteristic. Those tubules that are not sclerosed are lined by Sertoli cells, but these cells may be degenerate. Leydig's interstitial cells appear to be increased because the other elements of the testes are reduced in quantity. Spermatogenesis is usually entirely absent but some may be seen in a few tubules.

As one can see, apart from tiny testes in the adult, there are no constant clinical features of this syndrome. The chief problem of these patients is infertility. Some 40 per cent of aspermic males are examples of this syndrome.

One cannot do anything about the infertility but one may be able to do much to help these patients, even from early infancy.

The small penis, evidence of inadequate androgen output from the fetal testes, may be so diminutive that of itself it may be an anxiety to the parents making the baby unacceptable to them as a real boy. It can be brought to a normal size by early administration of testosterone and at puberty, again, it can be ensured that penile size, beard and pubic hair growth, masculinisation of the voice and male skeletal growth and muscle bulk are brought to an acceptable male phenotype.

Having already said that the shy, apprehensive, passive, and lonely personality is the greatest handicap of these patients we can now, I think, indicate that this handicap may be reversible with testosterone. With testosterone replacement therapy, and very quickly, the patient may become more outgoing, more sociable, more confident, happier and may give up his delinquencies, his crimes and, if such is his problem, his alcoholism. The late David Smith described one patient treated with testosterone who, when asked what was the most important difference he himself had noticed, replied, 'I now have friends'. In the treated patient those friends may be, for the first time, girl friends.

No one is quite sure at the present time at what age one should start testosterone. Probably one should give a burst of androgenic stimulation if the newborn penis is very small; but should one continue? I think the answer probably should be yes, for even in the preschool and school years the passive, unambitious, lonely personality may cause, at the least, unhappiness, and at the worst poor performance, behaviour problems and delinquency. It really seems that the patient with Klinefelter's syndrome may be as much in need of testosterone as the diabetic of insulin.

Cytogenetic Features of Klinefelter's Syndrome
In the usual case the chromosome complement is 47,XXY: an extra X is

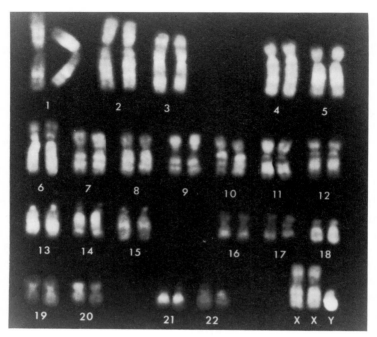

Fig. 128. Karyotype of 47,XXY Klinefelter's syndrome.

present in this undoubted male (Fig. 128). Just why this excess X should cause the changes that it does are far from clear but those changes are evidence that, although the presence of a Barr body indicates that one X is condensed and visible, there is not complete inactivation.

I have not seen it stated whether the extra X is maternally or paternally derived, or whether it might have either origin. One suspects the former since there is a slight, but only slight, relationship to maternal age.

In about 20 per cent of cases of Klinefelter's syndrome there is mosaicism: 46,XX/47,XXY/47,XXY.

It can happen that two, three, or even four extra X chromosomes are present. The tetra X (XXXX) plus Y complement and clinical picture are well recognised (Fig. 129).

There are likely to be features such as low birth weight and continuing growth failure, wide-set eyes with a 'mongoloid' slant, some limitation of joint movement at the elbow and, always, significant mental retardation. Published pictures of the XXXXY syndrome are remarkably like one another (see Figs 130 and 131).

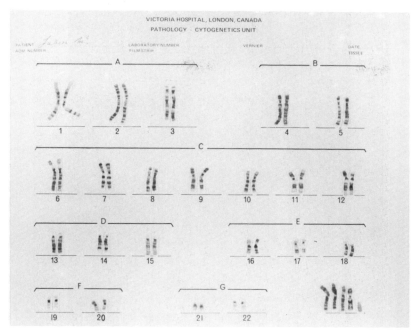

Fig. 129. Karyotype of tetra XY Klinefelter's syndrome, 48,XXXXY.

'Males, 46,XX'

Rather rarely a patient may be seen with a phenotype very similar to that of Klinefelter's syndrome and yet the chromosome complement is what one finds in a normal female: 46,XX. What might be the explanation? It is not clear.

It may be that there was originally a mosaicism, an XX/XY complement and that the XY stem-line of cells died out. It might be that a mosaicism still exists but the XY stem-line has not come to light. It could be that there has been a translocation so that there is hidden within one X chromosome the fragment of the short arm of Y that codes for the HY antigen. We just do not know.

The YY Complement, XYY

In 1961 a man with an XYY (Fig. 132) chromosome complement was dis-

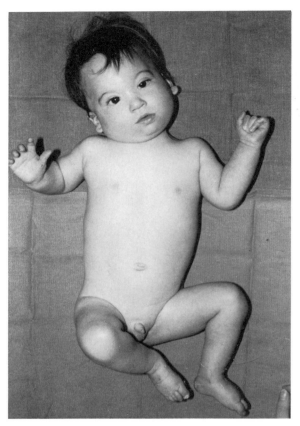

Fig. 130. Baby with 48,XXXXY complement; the appearance is rather characteristic.

covered almost by accident when he was investigated because he had a child with Down's syndrome. In 1967 much interest was aroused, especially by the news media, when Jacobs and her co-workers published an account of seven XYY men among 197 inmates of a Scottish institution for mentally subnormal males with aggressive and psychopathic behaviour. Jacobs noted that most of these XYY delinquent males were over 181 cm (72 inches) tall. The facts seemed to indicate that an extra Y chromosome was associated with aggressive delinquency though Hunter made the point that a large forbidding man might more likely be put in a security institution than a small man who had committed the same offence. Be that as it may, the Y chromosome achieved much notoriety as 'the criminal chromosome'.

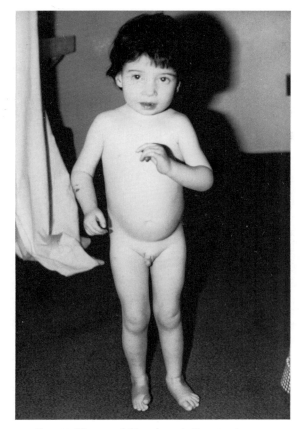

Fig. 131. The same child as shown in Fig. 130 at 4 years.

Since then a much more moderate view has come about for it has been recognised from large studies of unselected neonates that about one in 750 male babies born has an XYY complement. If all these were criminally and aggressively insane we would all be in fear and trembling for our lives! The truth is that the majority, indeed the great majority, are normal in our midst, unrecognised as unusual in infancy, childhood, adolescence and as adults (Figs 133, 134). On the average, they are not exceptionally tall.

But there are exceptions. Only today, the day of writing this section, I saw a 10-year-old boy with an XYY complement. At 130 cm and 53 kg he is enormously big for his age. He has a voracious appetite and will steal to get food. He cannot be disciplined in the home; he is loud-

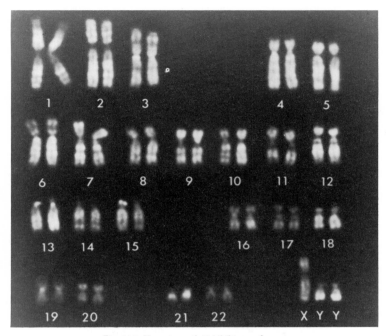

Fig. 132. Karyotype of 47,XYY complement.

mouthed, insolent and hyperkinetic. In school he has severe problems, especially in verbal comprehension. Apart from his huge size he shows no specifically unusual features except a slight degree of prognathos. Genital development is appropriate for a boy of 10. One can anticipate that he will reach at least 200 cm tall. One can anticipate, too, that he is destined for much trouble.

What can one say in general, then, about the XYY complement? The great majority are quite normal and live unremarkable lives. There may be some increase in problems of speech and language development in early childhood, but usually these problems are transient. For all we know, any of us, even myself, may have an XYY complement! But there are, for reasons that are not understood, rare exceptions. Whereas the incidence at birth is 0.13 per cent of all males, the prevalence in institutions for repetitively delinquent mentally sick patients may be about 2 per cent. Their crimes are not of aggression but of foolish irresponsibility: petty theft, vandalism and arson.

If one encounters by chance, let us say at an amniocentesis done for late maternal age, an XYY fetus, what should one say or do? It is very far

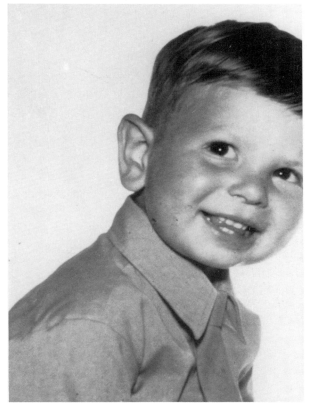

Fig. 133. Peter M, 47,XYY, aged 3 years; normal appearance.

from easy for one cannot know whether this baby will grow and progress according to the rule or to the exception. If one discloses what one has found, the parents may elect to abort what would in all probability have been a normal child with normal potential. If the pregnancy continues despite disclosure of the XYY complement and its possible, though unusual, significances one can imagine that every normal childhood peccadillo will be regarded as a sinister portent of worse to come. One can further imagine that upbringing in such an atmosphere might generate a self-fulfilling prophecy. But what if one says nothing, where would one stand with one's conscience—and the law—if this child turned out to be an incorrigible delinquent. One parent was most hostile when we told him that we had, in a neonatal survey, found that his child

Fig. 134. Peter M aged 10 years; no unusual features or behaviour but some problems of language development and reading requiring remedial teaching. Such problems are very common in children with normal chromosome complements.

had an XYY complement. He felt that we had no right to do the test and, furthermore, we had no right to tell him the result. How would he have felt, one wonders, if he learned that we had not told him what we had found? I think one has to tell what one has found. One rarely regrets telling the truth—even if one does not know just what to say!

The Fragile X, Marker X-linked Retardation, Martin–Bell Syndrome

It has long been recognised that males show an increased frequency of

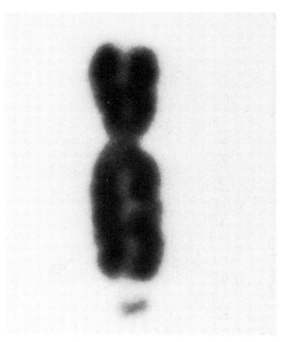

Fig. 135. The fragile X chromosome; the fragile site or constriction is at Xq27.3.

mental retardation, a 25 per cent preponderance in fact. The explanation offered, in the days of flourishing male chauvinism, is that there was ascertainment bias. The underachieving male was more likely to be noted than the defective female—from whom less, normally, could be expected!

In 1943 Martin and Bell had described a British family with 11 affected males in two generations. Similar families with many affected males were described by others. In 1962 Renpenning published a study of a large Canadian family containing 20 affected males with microcephaly. In 1972 the possibility that X-located genes might be responsible for the excess male preponderance was put forward by Lehrke, but was dismissed. But at least there was now a suggestion that there might be an X-linked form of mental retardation.

In 1969, Lubs of Denver, Colorado, investigating a family containing four males with apparent X-linked mental retardation found in the affected males a secondary constriction on the distal end of the long arm of the X chromosome, at a location now designated as Xq27.3 (Fig. 135).

He also found that an obligate carrier of the hypothesised X-linked trait also showed the secondary constriction that has now become known as the 'fragile X'. His findings were not reproduced by others and his discovery was forgotten until Giraud in 1976 and Harvey in 1977 found the same fragile X chromosome in some families with X-linked retardation.

In 1977 Sutherland made the most important observation that the composition of the culture medium was what determined if the fragile X would be seen or not. Only if the cells were grown in a culture medium deficient in folic acid, an old-fashioned medium, would the fragile X be seen. This was the one spectacular advance in cytogenetics referred to far back in this book (page 129).

Investigation of the family described by Martin and Bell using folic acid-deficient media showed that there was indeed this fragile X chromosome in the relevant members of the pedigree. Hence the suggestion that X-linked mental retardation should be called the Martin–Bell syndrome; however, this name has not gained general acceptance. Investigation of the family described by Renpenning did not show the fragile X. X-linked mental retardation with microcephaly but without fragile X is sometimes called Renpenning's syndrome. There are undoubtedly several varieties of X-linked mental retardation, of which the fragile X syndrome is only one.

The precise meaning of this secondary constriction, this fragile site, is unknown. Is it that there is, in this specific location, a gene (or gene deletion) that causes the mental retardation, and that the genetic effect is more evident in the hemizygous male than the heterozygous female? Or is it that there is an X-linked gene that causes the mental retardation but also causes the secondary constriction as a cell-culture peculiarity, an epiphenomenon, which could be part of a much more general phenotypic effect. Perhaps the latter is the more likely, for the fragile X chromosome is but a cultural artefact; there is no saying that its presence and the syndrome are cause and effect.

While almost all, but not quite all, males with the fragile X show mental retardation varying from moderate to severe, there may be enough other features to constitute a dysmorphic syndrome. The birth weight is above average, but the adult height is not. The forehead is large, the nose long, the ears large, the chin prominent and the whole face elongated (Fig. 136). Some patients have grand mal seizures.

Testicular enlargement is almost universal in the postpubertal male with the fragile X. Whereas the upper limit of testicular volume for Caucasian males (it is smaller in Asiatic races) is 25 ml for each testis, the tes-

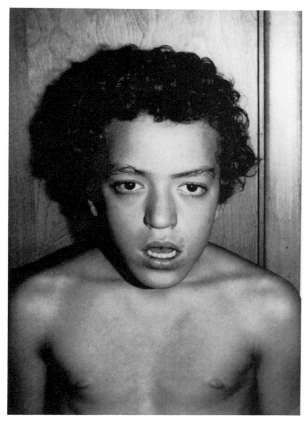

Fig. 136. *Typical facial features associated with the fragile X chromosome in a male; this phenotype is not invariably present.*

ticular volume in the syndrome may be up to 60 ml. The formula for determining testicular volume is length × width^2 × $\pi/6$. Very recent reports indicate that the testes may be abnormal in prepubertal males (Fig. 137) and even in early infancy. The histology of these macro-testes is quite non-specific. There may be some increase in either interstitial fluid or supporting tissue, but there is no reason to suppose that they do not function normally.

The personalities of these patients is said to be pleasant and jocular, but with often quite severe speech defects. Characteristically there is staccato, jerky speech with 'litany language'—the repetition of words or

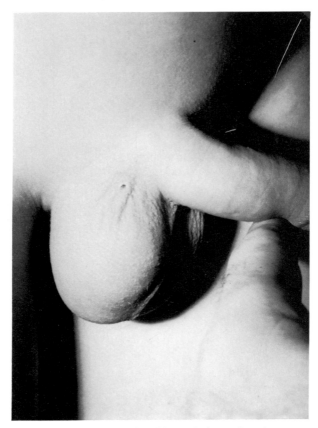

Fig. 137. Macro-orchidism in a prepubertal boy with the fragile X chromosome; macro-orchidism has been recognised in a young baby with the fragile X.

phrases. The IQ ranges between 30 and 80, with average about 60; there have been one or two reports of males with normal intelligence.

What of the females? This is difficult. Even in the obligate carriers of the fragile X (the mother of one or more affected males) there are no physical phenotypic features. About 30 per cent have some degree of mental impairment which, at the worst, is rarely more than mild. To look at it in another way: Turner in Australia investigating girls with mild mental retardation found that 7 per cent showed the fragile X. Undoubtedly the fragile X is associated with a rather high risk of mental retardation in females. The gene, whatever it is, is not always entirely without phenotypic effect. That may of course be that because in women showing mental retardation, uneven lyonisation has led to the majority of cells having in the same active state the X chromosome carrying the fragile site and the abnormal genetic function.

The difficulty, then, lies in the fact that many obligate carrier females do not show the fragile X. Whether the peculiarity of the X chromosome can be demonstrated or not seems to depend on age and mental status. An obligate carrier who is young and definitely retarded will very probably show the fragile X. A young woman who is not retarded probably will not. An elderly retarded woman might show it on very careful searching; an elderly but not retarded woman almost certainly will not. Even at best, the carrier female shows very few fragile X-positive cells. Whereas the affected male may show up to 50 per cent positive cultured lymphocytes, an obligate carrier female may show no more than 4 per cent of positive cells. It can be very difficult to decide on cytogenetic grounds who, or who is not, a carrier.

If the fragile X chromosome is to be found in either affected males or suspect carrier females the laboratory must be informed of the purpose of the chromosome study so that they may know to use a folic acid-free medium or even try to enhance the fragile X effect by incorporating in the culture medium a folic acid antagonist such as FUdR or methotrexate.

Does all this matter? Is it merely of academic importance, an exercise in cytogenetic subtleties? Far from it. It is reckoned that at least one in every 2000 males has fragile X mental retardation, and it may be nearer one in 1000. Fragile X mental retardation is, after Down's syndrome, the second most common single cause of mental retardation.

If we look at an actual pedigree of a family that recently came to my notice (Fig. 138) we can see that if the fragile X had been recognised as being present in this family, the birth of the retarded males Robert, Mark and Daniel might have been prevented by suitable genetic counselling and prenatal diagnosis—if there had been knowledge of these matters in those days.

If it is known that a woman is a carrier of the fragile X, any son will be at a 50 per cent risk of being retarded. Any daughter will be at a 50 per cent risk of having received the fragile X. If she has, she will be at a 30 per cent likelihood of being slightly retarded.

If nothing else, the known carrier female could be offered prenatal sex determination of the fetus, either by amniocentesis at 16 weeks' gestation or by chorion villus sampling at 9 weeks. It might be justifiable to offer abortion of a male fetus, as it has a 50 per cent chance of handicapping retardation; a female fetus has only a 15 per cent (30 per cent of

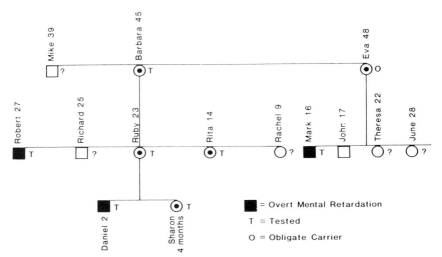

Fig. 138. *Pedigree of family with the fragile X; the female carriers of the abnormal chromosome are not retarded. It remains to be seen whether baby Sharon will be retarded. At the worst she will not be severely defective being heterozygous for the fragile X chromosome.*

50 per cent) of slight intellectual impairment. Can we do better? It is possible.

Amniotic fluid cells, cultured in a medium containing a folic acid antagonist may show the fragile X in a potentially affected male, but probably would not even in a carrier female. Much work still remains to be done on prenatal diagnosis.

However, there is a new hope. Almost inevitably that hope lies in recombinant DNA technology, marker DNA sequences and gene-linkage. At the present time at least two marker restriction fragment polymorphisms are known close, but not close enough, to the fragile site, Xq27.3. The sequences presently identifiable by specific DNA probes are not close enough so there is little likelihood of meiotic recombination (Fig. 34) separating the marker sequences from the fragile X site. Closer sequences or an intragenic and specific probe will surely soon be found. There is indeed good hope that the considerable burden to society and to individual families of fragile X mental retardation can be reduced.

Recommended Further Reading
The Fragile X

Turner G. and Jacobs P. (1983). 'Marker (X)-linked mental retardation'. *Advances in Human Genetics, Vol. 13*, pp. 83–112. Plenum Press, New York and London.

CHAPTER 13

The Intersex States

Hermaphroditism and Pseudohermaphroditism

The essential feature of true hermaphroditism is that both types of gonads, male and female, are represented. There may be, most commonly, an ovary on one side and a testis on the other; or there may be a mixed gonad, an ovotestis, on one side and a testis or ovary on the other. Rarely there may be ovotestes on both sides. Cytogenetically there may be a mosaicism: 46,XX/46,XY or 45,X/46,XY but there may be just a single stem-line-detectable 46,XX or 46,XY. In these cases one can only presume that there has been (and still may be) a mosaicism in the past, or that there has been an error of genetic instruction rather than a chromosome disorder.

In the majority of cases of true hermaphroditism the testicular element predominates in function. There is significant, if only partial, masculinisation. Hermaphrodite babies are usually, at first sight, designated as males.

Pseudohermaphrodites may be either male or female: uniformly male or uniformly female gonads respectively, but both with genitalia that may at the least be ambiguous and at the most frankly discordant with the gonadal sex.

Male pseudohermaphroditism is rarely due to a chromosome anomaly. If it is, it is likely that there will be a mosaicism, 46,XY/45,X. In this condition the genitalia are likely to be sufficiently masculine that the baby will be declared a boy. At puberty there will probably be sufficient testicular hormonal function for there to be an increase in penis size, pubic hair and deepening of the voice. But the testes, functionally active though they may be, are dysgenetic and at real risk of malignant change. They should be removed.

Some cases of male pseudohermaphroditism result from more or less complete failure of the germinal ridge to develop as testes for other than chromosomal reasons. There may be more or less failure of coding for the HY antigen from genetic failure of the Y chromosome or its genetic colleagues. The gonads, such as they are, will be male, and there will be more or less male gonadal dysgenesis. At the worst there may be no more than 'streak testes'. These dysgenetic gonads, too, are at risk of malignant change and should be removed. The internal genitalia may be entirely female, for there may be insufficient Mullerian duct repression factor to inhibit their development; or there may be, internally, features of both male and female organs—'pure' or 'mixed' gonadal dysgenesis respectively.

There are cases, admittedly rare, of male pseudohermaphroditism resulting from more or less failure to produce testosterone because of a metabolic block in one of the five steps of testosterone synthesis (see Fig. 109). There are also, just as rare, cases of male pseudohermaphroditism as the result of deficiency, as an autosomal recessive inborn metabolic error, of the enzyme 5-alpha reductase (see Fig. 109). In either case the chromosome complement will be 46,XY.

Most commonly and most importantly male pseudohermaphroditism is due to more or less failure of the embryonic genital cells to take up dihydrotestosterone and to react to its instructions. If the failure of receptor activity is complete, there will be the classical testicular feminisation syndrome. If receptor activity is merely incomplete one will see partial testicular feminisation syndrome (see Fig. 123). If there is nearly normal response to dihydrotestosterone there will be nearly normal male genitalia, Reifenstein syndrome. The complement will be, of course, 46,XY.

Female pseudohermaphroditism, 46,XX, with ovarian tissue has only one cause: excess androgenic hormones affecting the external genitalia of the fetus. This could come about, indeed in the past did come about, by maternal ingestion of progestational drugs with an androgenic effect. It could come about if the mother had, herself, an androgen-secreting adrenal or ovarian tumour. By far the largest number of cases of female pseudohermaphroditism are due to virilising adrenal hyperplasia, one of a number of inborn errors of metabolism affecting the synthesis of cortisol (see Figs 110, 111).

Ambiguous Genitalia

There are few more difficult situations facing a clinician than when, at the moment of birth, he declares 'It's a ... a ... a ...'. The genitalia are ambiguous. Society expects an instant declaration; it is baffled by an ambiguity. There are males; there are females. There are no 'its'.

One cannot tell, by looking at the external genitalia, whether this baby is a true hermaphrodite, a male pseudohermaphrodite or a female pseudohermaphrodite. They all look the same. One cannot tell by mere inspection whether this baby is an 'unfinished' male or a partly virilised female.

I have found the term 'unfinished' to be a good one. It is much less distressing to parents than such a term as 'half and half'. It allows of a simple explanation of the embryology of the genitalia and how nature's failure to finish the transformation to a male can leave the external genitalia more or less ambiguous.

When faced with the immediate dilemma one can only tell the parents that the genital organs have not developed correctly and that some thought must be given as to how the problem is to be handled. If pressed to give a decision as to sex assignment, naming and birth notices one should, I believe, designate the baby as a female—unless one is very sure that the phallus could be surgically constructed to be acceptable as a penis in the locker room and functionally adequate for urination standing. It is very, very difficult (but surgical optimism will deny this) to construct a good penis from ambiguous genitalia. I have seen pathetic penises considered surgical successes. A phallus, if unlikely to meet the standards demanded by society, can be removed and the patient will be unembarrassed among his peer male group. He, now she, can sit like a lady. It is most unkind to perpetuate or create an inadequate male merely for the sake of genetic tidiness.

A chromosome study of course is of great value. If the complement is XX, this may well be a case of virilising adrenal hyperplasia (see Fig. 111), and one must be alert to the possibility of an acute crisis of adrenal insufficiency. But with an XX chromosome complement there might be true hermaphroditism, though this is much more rare than female pseudohermaphroditism.

If the chromosome complement is 46,XY one might be looking at a case of true hermaphroditism, male gonadal dysgenesis or a case of incomplete testicular feminisation syndrome.

It is nice, tidy and intellectually satisfying to have a 'scientific' explanation of the ambiguous genitalia. But that matters much less than the happy adjustment in society of this unfortunate newborn.

People can, and do, live happily without sex or reproduction, but in our society one goes, as one must, to either the 'Gents' or the 'Ladies'. In dealing with ambiguous genitalia one must remember, unless there are most pressing reasons to think otherwise, 'The call of nature is the moment of truth'.

CHAPTER 14

When to Order a Chromosome Study

Needless to say, no laboratory investigation should be ordered frivo-lously or merely 'for the sake of completeness'. There must be a good reason. Apart from the discomfort and anxiety they may cause, they cost money. The funds available for health care are not limitless; money wasted on unnecessary tests leaves less money for more worthwhile matters. Chromosome studies are exacting, time-consuming and expens-ive. A blood lymphocyte study costs about $150 (Canada, 1985). There is little left from $500 for prenatal diagnosis by amniocentesis. When, then, is such an expenditure justified?

In a Baby, Young Child or Adult Patient

Even if, on clinical grounds, the diagnosis of a chromosome disorder seems certain, a karyotype analysis is in order. The diagnosis should be confirmed. There can be surprises. The diagnosis of a chromosome abnormality usually has important implications and it is as well to sup-port clinical opinion with laboratory confirmation. It can, moreover, be very important to know the genetic mechanism responsible for the abnormality; has the Down's syndrome come about by the chance mis-hap of non-disjunction or because one or other parent carries a balanced translocation? Or has it come about by isochromosome formation? Is it an example of mosaicism? The risk of Down's syndrome in subsequent children is much related to the answers to these questions, as it may be for other chromosome disorders.

Mental retardation alone, without any specific dysmorphic features or a pedigree suggesting X-linked retardation, is quite unlikely to be asso-ciated with a chromosome anomaly, and a chromosome study is hardly justified in non-specific retardation. But if there are dysmorphic features, if the patient is (in a term often used by paediatricians and geneticists) an FLK or 'funny-looking kid', along with the retardation, a careful study should be done by a banding technique. It might reveal a small deletion or addition of genetic material that might have come about because a parent carried a balanced translocation. Remember that unless the case is typical of a recognised chromosome anomaly, unless only confirmation of a diagnosis is required the laboratory should be alerted to the fact that the case is not typical and that small additions or deletions should be sought.

If there appears to be X-linked mental retardation, or if there is macro-orchidism or a suggestive facies (see Fig. 136) a chromosome study is certainly justified, but the laboratory must be informed of the nature of the problem. They must know that the cells must be cultured in a medium free of folic acid, otherwise the test is useless.

If a neonate has ambiguous genitalia, a chromosome study may help to define the mechanism that led to the abnormality, though the knowledge gained of the chromosomal and genetic sex must not necessarily influence the gender assignment and the sex of rearing of the child.

Primary amenorrhoea and failure of pubertal development in a female warrant a karyotype. She might have a 45,X chromosome complement; or she might have an XY complement and be an example of the testicular feminisation syndrome. A female sib of a known case of that syndrome should also have chromosome studies, as she, too, might have the syndrome of androgen unresponsiveness.

Male infertility with aspermia or extreme oligospermia should have a chromosome study. He may have no overt features of Klinefelter's syndrome but may have an XXY complement. If nothing else, one might have an explanation of the infertility.

One must remember that not all girls with a 45,X complement show obvious features of Turner's syndrome. A girl with unexplained short stature deserves a chromosome study, as does perhaps any girl with coarctation of the aorta or Meckel's diverticulum. These abnormalities are uncommon in 46,XX females.

In the Parents of a Patient

If the patient, the 'index case' shows a translocation chromosome anomaly one will want to examine the karyotypes of both parents. If their chromosomes are normal, the abnormality arose in their child *de novo* as a unique event. Other children would be at negligible risk of a similar abnormality. But if one or other parent carries a balanced chromosome translocation there is a significant risk of repetition with other children; prenatal diagnosis should be offered in another pregnancy.

A child with a trisomy not due to translocation but to non-disjunction, or a child with a disorder due to isochromosome formation or to mosaicism, does not justify chromosome studies of the parents, as they will be normal.

Habitual Spontaneous Abortion

Spontaneous abortion, miscarriage, is a common event. About 15 per cent of all recognised pregnancies end in miscarriage and, to be sure, about 50 per cent of those many miscarriages are of a fetus with a chromosome anomaly; but almost always the anomaly has arisen in that fetus as a *de novo* event, a chromosome mutation, without any abnormality of the parental chromosomes. A single miscarriage—or even two—does not justify chromosome studies of the parents.

However, if there have been three or more spontaneous miscarriages the term 'habitual abortion' becomes applicable. Habitual abortion can have many causes, uterine malformation, smouldering chlamydia infection of the uterus, even immunological rejection of the fetus, for example. But it might be that one or other of the couple carries a balanced chromosome translocation and that successive fetuses are being recognised by the 'quality control' of the uterus as having unbalanced chromosome complements. The yield of abnormal results from couples with habitual abortions is not great: 5 per cent, or thereabouts, in couples with three successive miscarriages would not be far out. It can be worth a look.

The Fetus: Prenatal Chromosome Analysis

Whether the technique is the standard method of amniocentesis or the less well-tried chorion villus sampling, there are at times good reasons to offer prenatal chromosome analysis.

As we have seen (see Fig. 53) the risk of any non-disjunction chromosome trisomy increases with advancing maternal age. One cannot draw a firm line between an age that justifies prenatal diagnosis and one that does not, but most clinics encourage chromosome studies of the fetus at a maternal age of 35 and greater.

Because, for reasons that are not fully understood, a child with any chromosome anomaly (except perhaps *de novo* deletion, ring chromosome, isochromosome and mosaicism) is more likely to be followed by another child with a (not necessarily identical) chromosome abnormality, an affected child justifies an offer of prenatal diagnosis in another pregnancy, even if the parents' chromosomes are known to be normal. One cannot make a strong case for such prenatal diagnosis, but the offer should be made.

Of course, if one or other parent is known to be the carrier of a balanced translocation, prenatal diagnosis should be offered. The risk of a fetus with an unbalanced complement is high, though by no means so high as theory would suggest.

Finally, all that one may be able to offer, but should offer in certain cases, is prenatal sex determination: is the fetus XX or XY? If there is an X-linked and serious disease that cannot definitively be diagnosed prenatally (Duchenne muscular dystrophy in June 1985) sex determination can be most relevant. An XX fetus would not have the disease; an XY fetus would be at a 50 per cent risk of having the serious X-linked disorder. If that disorder is bad enough—and if there is no effective treatment—'therapeutic' abortion of all XY fetuses could be condoned.

Stop press news! There are unconfirmed rumours that a definitive probe for the gene for Duchenne dystrophy has been devised (May, 1986). We must wait and see.

Index

215